# Extending the S System

# The Wadsworth Statistics/Probability Series

**Series Editors**

Peter J. Bickel, University of California, Berkeley
William S. Cleveland, AT&T Bell Laboratories
Richard M. Dudley, Massachusetts Institute of Technology

Richard A. Becker and John M. Chambers
*Extending the S System*

Richard A. Becker and John M. Chambers
*S: An Interactive Environment for Data Analysis and Graphics*

Peter J. Bickel, Kjell Doksum, and J. L. Hodges, Jr.
*Festschrift for Erich L. Lehmann*

George E. P. Box
*The Collected Works of George E. P. Box*
Volumes I and II, edited by George C. Tiao

Leo Breiman, Jerome H. Friedman, Richard A. Olshen, and Charles J. Stone
*Classification and Regression Trees*

John M. Chambers, William S. Cleveland, Beat Kleiner, and Paul A. Tukey
*Graphical Methods for Data Analysis*

Franklin A. Graybill
*Matrices with Applications in Statistics*, Second Edition

John W. Tukey
*The Collected Works of John W. Tukey*
Volume I: *Time Series, 1949–1964*, edited by David R. Brillinger
Volume II: *Time Series, 1965–1984*, edited by David R. Brillinger

# Extending the S System

Richard A. Becker
John M. Chambers

WADSWORTH ADVANCED BOOKS AND SOFTWARE

Monterey, California

Learning Resources
Centre

Wadsworth Advanced Book Program
A Division of Wadsworth, Inc.

This book was typeset in Palatino by the authors.

Printed in the United States of America
 2 3 4 5 6 7 8 9 10 — 89 88 87 86

**Library of Congress Cataloging in Publication Data**

Becker, Richard A.
  Extending the S system.

  Includes index.
  1. S (Computer system)   2. Interactive computer
systems.   3. Statistics—Data processing.
4. Mathematical statistics—Data processing.
I. Chambers, John M.   II. Title.
QA276.4.B424   1985      001.64'2        84-29933
ISBN 0-534-05016-6

# Contents

Introduction . . . . . . . . . . . . . . . . 1

## 1 Writing New Functions for S . . . . . . . . . . . 3

1.1 Design and Implementation of Simple S Functions . . 4
1.1.1 Design of S Functions . . . . . . . . . 4
1.1.2 Arguments; the FUNCTION Statement . . . . . 6
1.1.3 Error Checking . . . . . . . . . . . 7
1.1.4 Dynamic and Static Data Structures . . . . . . 8
1.1.5 Computations . . . . . . . . . . . 10
1.1.6 Creating New Functions . . . , . . . . . 11
1.1.7 Hints; Debugging . . . . . . . . . . 14
1.1.8 Strange Happenings; Warts . . . . . . . . 16

1.2 Data Structures . . . . . . . . . . . 17
1.2.1 Data Values and Attributes . . . . . . . 17
1.2.2 Character Data and Character Attributes . . . . 18
1.2.3 Missing Values . . . . . . . . . . 21
1.2.4 Structures and Components . . . . . . . 21
1.2.5 Modes Computed at Execution . . . . . . 23

1.3 Function Arguments . . . . . . . . . . 27
1.3.1 Arguments in the FUNCTION Statement . . . . 28
1.3.2 Interrupting and Resuming Argument Processing . 29
1.3.3 Arbitrarily Many Arguments . . . . . . . 30
1.3.4 Treating Structures Like Argument Lists . . . . 32

1.4 Function Results and Related Statements . . . . 34
1.4.1 The RETURN Statement . . . . . . . 34
1.4.2 CHAIN: Invoking Another Function . . . . . 36

1.4.3  INSERT: Building Structures . . . . . . . . 37

1.5  Graphics Functions . . . . . . . . . . . . . 37
1.5.1  Declaring a Graphics Function . . . . . . . 37
1.5.2  Graphical Parameters . . . . . . . . . . . 37
1.5.3  Plotting Data Structure . . . . . . . . . . . 39
1.5.4  High Level Graphics: SETUP and LOGPLOT . . . 39

**2  Writing and Using Algorithms** . . . . . . . . . . **41**

2.1  The Algorithm Language: Basics . . . . . . . . 41
2.1.1  Languages . . . . . . . . . . . . . . . 41
2.1.2  MAKE: Generating S Functions . . . . . . . 42
2.1.3  Error Handling . . . . . . . . . . . . . 43
2.1.4  Symbolic Constants; Declarations . . . . . . 43
2.1.5  Debugging . . . . . . . . . . . . . . . 45
2.1.6  C Language Facilities . . . . . . . . . . . 45

2.2  Special Facilities . . . . . . . . . . . . . . 46
2.2.1  Printing and Encoding . . . . . . . . . . 47
2.2.1.1  Basic Message Printing . . . . . . . . . 47
2.2.1.2  Encoding . . . . . . . . . . . . . . . 49
2.2.1.3  Detailed Format Control . . . . . . . . . 50
2.2.1.4  Problems with Encoding . . . . . . . . . 51
2.2.2  Reading and Decoding . . . . . . . . . . 52
2.2.2.1  Basic Reading of Data Items . . . . . . . 52
2.2.2.2  Line Input; End-of-File . . . . . . . . . 54
2.2.3  File Access; Standard Files . . . . . . . . . 55
2.2.4  Dynamic Storage . . . . . . . . . . . . . 56
2.2.5  Data Structures for S . . . . . . . . . . . 57

2.3  Available Algorithms . . . . . . . . . . . . 59
2.3.1  Data Handling; Character Data . . . . . . . 60
2.3.2  Sort and Order . . . . . . . . . . . . . 61
2.3.3  Range of Data . . . . . . . . . . . . . . 62
2.3.4  Probabilities; Quantiles; Pseudorandom Numbers . . 63
2.3.4.1  Available Algorithms . . . . . . . . . . 63
2.3.4.2  New Pseudorandom Generators for S . . . . . 63
2.3.5  Matrices and Arrays . . . . . . . . . . . 64

**3  Graphical Algorithms** . . . . . . . . . . . . **67**

3.1  Basic Concepts of Graphical Algorithms . . . . . 68
3.1.1  Figures and Plots . . . . . . . . . . . . 68
3.1.2  User and Margin Coordinate Systems . . . . . 68
3.1.3  Graphical Parameters . . . . . . . . . . . 70

3.2  Creating Graphical Algorithms . . . . . . . . . 71
3.2.1  Initialization of Graphical Algorithms . . . . . 71
3.2.2  Setting-up Coordinate Systems and Axes . . . . 72
3.2.3  Drawing the Picture . . . . . . . . . . . . 73
3.2.4  Titles and Axis Labels . . . . . . . . . . 74
3.2.5  Specifying and Querying Graphical Parameters . . 74
3.2.6  Wrapping Up . . . . . . . . . . . . . . 77

3.3  The Structure of a Graphical Algorithm . . . . . . 78
3.3.1  High-Level Graphical Algorithms; S Functions . . . 78
3.3.2  Algorithms that Augment a Plot . . . . . . . 82

3.4  Advanced Graphical Algorithms . . . . . . . . 83
3.4.1  Control of Figures, Plots, Margins . . . . . . 83
3.4.2  Margins and Outer Margins . . . . . . . . 86
3.4.3  Parameters of Physical Size . . . . . . . . 88
3.4.4  Setting-up Coordinate Systems and Axes . . . . 89
3.4.5  Summary of Graphical Parameters . . . . . . 91
3.4.6  Graphical Input . . . . . . . . . . . . 93
3.4.7  Debugging . . . . . . . . . . . . . . 94

3.5  Available Graphical Subroutines . . . . . . . . 94

3.6  Device Drivers . . . . . . . . . . . . . 96
3.6.1  Organization of Device Driver Routines . . . . 97
3.6.2  Portability Considerations . . . . . . . . 102
3.6.3  Control of Graphic Input . . . . . . . . . 102
3.6.4  Writing a Device Driver for S . . . . . . . 103

Appendix A  Interface and Algorithm Languages . . . . . 105

A.1  Interface and Algorithm Languages . . . . . . 105
A.2  Grammar for Interface Language . . . . . . . 117

Appendix B  Definition of the S Language . . . . . . . 121

B.1  The Language: Syntax . . . . . . . . . . . 121
B.1.1  Function Calls and Commands . . . . . . . 125
B.1.2  Compound Expressions . . . . . . . . . 125
B.1.3  Continuation . . . . . . . . . . . . . 125
B.1.4  Reserved words . . . . . . . . . . . . 126

B.2  Semantics . . . . . . . . . . . . . . . 126
B.2.1  Functions and Operators . . . . . . . . . 126
B.2.2  Side Effects: Database Changes and Parameters . . 130
B.2.3  Compound Expressions . . . . . . . . . 131
B.2.4  Conditional Expressions; Iterative Expressions . . 132

B.3  Data Structures . . . . . . . . . . . . . 133

**Appendix C  Documentation for S Utilities** . . . . . . . 137

**Appendix D  S Manual Pages** . . . . . . . . . . . 145

**Appendix E  Maintaining S** . . . . . . . . . . . . 149

**Index** . . . . . . . . . . . . . . . . . . 155

# Extending the S System

# Introduction

One of the helpful features of the S system for data analysis and graphics is that users are encouraged to *extend* the system to suit their special needs. The S macro facility is one approach to extending S, as also are the use of the S *menu* function and of the REPORT facility. A different way to extend S is to write *new* functions which make use of algorithms to carry out computational tasks. The extensions possible by this route are unlimited, and often produce more efficient solutions for users' problems. By this method, S can be merged with arbitrary computational techniques, even when the techniques were not designed originally to be used with S at all. As new methods, improved algorithms, and new areas of application arise, new functions can incorporate the advances into the S language. For users interested in designing an algorithm to solve a computational problem, working within the S environment, with a new S function that calls the new algorithm, is often much more effective than conventional programming environments. S provides easy ways to generate test data, manage results and do printing and plotting, all of which can otherwise be time-consuming tasks during program development.

This book describes how to extend the capabilities of S by writing new functions. It assumes that the reader has reasonable familiarity with the UNIX* operating system and some experience in programming.

The *interface language* is the key to defining new functions. Each S function is defined by an interface routine, written in the special-purpose interface language. When a user invokes the function, the interface program interprets arguments, carries out computations, and returns the results. Chapter 1 describes the S interface language.

---

* UNIX is a trademark of AT&T Bell Laboratories.

The computational job of an S function is usually carried out by algorithms called from the interface routine. S provides an environment of software support for the writing of such algorithms. Chapter 2 describes this environment, which centers on an *algorithm language*. Chapter 3 describes the S device-independent graphics facility, and shows how to write graphical functions and algorithms, as well as new device drivers. Appendix A gives a tabular description of the S interface and algorithm languages. Appendix B describes, in a formal way, the S language. Appendix C contains documentation for a number of S utilities that are used when writing new functions. Appendix D is a set of manual pages which describe S and some routines for constructing S datasets. Appendix E is designed to assist people who are charged with maintaining S.

This book extends the description of the S system in the book *S: An Interactive Environment for Data Analysis and Graphics*, (Richard A. Becker and John M. Chambers, Wadsworth, 1984), which we will refer to as Becker and Chambers (1984) in the present book. The reader should have a copy of this at hand, along with a good reference on the UNIX operating system. Programming aspects of the UNIX operating system are covered in *The UNIX Programming Environment* (Brian W. Kernighan and Rob Pike, Prentice-Hall, 1983).

# 1
# Writing New Functions for S

The S macro facility (described in Chapter 6 of Becker and Chambers (1984)) allows users to extend the S language by defining macros (which usually have one or more arguments) to carry out specialized computations. Eventually the macros must expand into expressions in the S language. This somewhat limits their efficiency and generality, since there are certain computations that are difficult to carry out using existing S functions. A different way to extend the language is to write *new* functions in S that typically make use of algorithms (subprograms) to carry out some computational task. As new methods, improved algorithms, and new areas of application arise, new functions incorporate the advances into the S language. The *interface language* is the key to defining new functions. Each S function is defined by an interface routine, written in the special-purpose interface language. When an S user invokes the function, the interface program is responsible for interpreting the arguments to the function and for returning the result of the function.

The computational job of the function is usually carried out by algorithms called from the interface routine. In this section, we will treat the underlying algorithms as given. They may be FORTRAN routines taken from a subroutine library or they may have been written for some other purpose, perhaps outside of S. In fact, S provides an environment of software support for the writing of such algorithms. Chapter 2 will describe this environment, which centers on an *algorithm language*.

Macros and new functions are both useful ways to extend the S language, but are useful in different circumstances. Macros build on the language by condensing complicated expressions into a single, simpler macro call. The user of the macro is relieved of the details of the underlying S expressions. When similar, but not identical, expressions recur frequently, the variations from one time to the next can often be neatly absorbed into the arguments to the macro. Chapter 6 of Becker and Chambers (1984) discusses the writing of macros. Macros, of course, can only produce computations in terms of existing S functions.

Sometimes the computation to be defined uses code that already exists in an algorithm (i.e., a subprogram) in, say, FORTRAN. Such an algorithm cannot be invoked directly from a macro. On other occasions, the computations may turn out to be inefficient when written in an obvious way in S. This particularly tends to be true when the computation involves iterating over single elements of the data. In these cases, the definition of a new S function, through writing an interface routine, may be the best approach. Writing the interface routine, collecting needed algorithms (or writing them), and getting the new function to work, involves a greater initial investment of time than creating a macro. It also assumes some experience on the writer's part with programming languages. Specifically, both the interface language and the *algorithm language* (to be described in Chapter 2) use RATFOR and the FORTRAN language as intermediate forms. Some familiarity with these is desirable, before one begins writing interface routines. For example, *Software Tools* (Brian W. Kernighan and P. J. Plauger, Addison-Wesley, 1976) gives many examples of Ratfor programs.

The essential principle in writing new S functions is: KEEP IT SIMPLE. Particularly for initial efforts, a clearly defined, well understood function will give best results, even though it may lack generality. It is more productive to use and adapt a simple function in the interactive environment of S than to spend time trying to debug an overly elaborate, poorly understood interface routine or algorithm.

## 1.1 Design and Implementation of Simple S Functions

This section works through a simple example to illustrate the use of the interface language. Imitating the style of this example will be a good way to get started in writing S functions.

### 1.1.1 *Design of S Functions*

Before doing any actual programming, one should have a fairly clear idea of the purpose and form of the new S function. What input information (arguments) it should use, what output information (results) it should provide and what computations it must do in between? If these are understood in general terms, and if they are clear and well defined, the implementation of the ideas in a new S function will be much easier.

We will consider a simple but typical example of an S function. Suppose we want to fit a linear model

$$Y = X \beta$$

to a matrix of data $X$ and a vector $Y$, given some existing algorithm(s) which carry out the desired numerical computations. We decide to call the function *fit*. Without worrying about details, here are some design decisions needed.

**Arguments**   We decide to fit a single variable to several other variables. As usual in S, the predicting variables are assumed to be columns of a matrix. We should check that the arguments make sense (for example, that they correspond to the same number of observations); otherwise the algorithm may return nonsense or fail for no apparent reason.

**Results**   A linear model, once computed, can be represented by the coefficients computed. However, it is usually more convenient for the user and more accurate computationally to compute the residuals as well. All the S functions treating linear models provide this information.

**Options**   Initially, we should try not to provide too many options. As experience with the new function grows, the important variations will become clearer. For linear models, two common options are: whether the model should include a constant (intercept) term; and the provision of weights for the observations (assuming an algorithm exists which handles weights).

**Algorithms**   We need to consider the algorithm(s) available, both to decide whether they are suitable as they stand and to make sure we provide all their requirements. For example, many computational algorithms require some working (scratch) storage. The interface routine provides this.

Throughout section 1.1 we will construct interface routines that all elaborate on the following example:

```
FUNCTION fit(
     x/MATRIX/
     y/REAL/
     )
if(NROW(x)!=LENGTH(y))
     FATAL(Number of observations in x and y must match)
STRUCTURE(
     coef/REAL,NROW(x)/
     resid/LIKE(y)/
     )
call lsfit(x,NROW(x),NCOL(x),y,coef,resid)
RETURN(coef,resid)
END
```

The remaining subsections of section 1.1 introduce in turn the main pieces needed to assemble an interface routine.

**1.1.2** *Arguments; the FUNCTION Statement*

The first statement of an interface routine is of the form:

FUNCTION name( $arg_1$, $arg_2$, ... )

where *name* is the name of the S function and $arg_1$, etc. are the descriptions of its arguments. Each $arg_i$ is defined by the argument name, followed by a *description list*:

name/ $spec_1$, $spec_2$, ... /

The description lists are used in the same form for arguments and for dynamically allocated data structures (section 1.1.4). For arguments, the specifiers $spec_1$, etc. give the desired type (MATRIX, TS, etc.) and mode (REAL, LOGICAL, etc.) for the arguments. The process of turning the actual arguments into the desired type and mode is called *coercion*. Details of the process will be introduced as we encounter more specialized needs. In the simplest form of *fit*, we want a vector of values, named *y*, and a matrix, *x*, to predict the vector:

FUNCTION fit( x/MATRIX/, y/REAL/ )

The example shows that some abbreviated specifications are possible. Those of most frequent use are:

- if the structure type is omitted, it is assumed to be a vector structure [Becker and Chambers (1984), section 5.1.2], so that the argument may be treated as a simple vector;
- if the type is given (e.g., MATRIX) but the mode is not, the default mode is REAL.

In the *fit* example, suppose that an argument, *wt*, provides optional weights for use in the fitting. If there exist two algorithms, one with and one without weights, the interface routine can simply check whether weights were provided and call the corresponding algorithm. We just need to make the argument *wt* optional, without providing a default data structure. When an argument is optional, the corresponding description list may say OPTIONAL, or may specify the default data structure to be allocated when the argument is missing. In the second case, the description is the same as would be used to allocate the structure dynamically (see section 1.1.4). In either case, the interface routine can make use of the logical expression MISSING(name) to test later whether the argument was missing in the call to the function.

We can also have an optional argument, *int*, to control whether the model should include an intercept term. In this case, the default action is to interpret the argument as TRUE. This can be built into the argument declaration, by specifying the default structure to be of length 1 and value TRUE:

```
FUNCTION fit(
    x      /MATRIX/
    y      /REAL/
    wt     /REAL,OPTIONAL/
    int    /LGL,1,TRUE/
    )
```

We see also in the above that the FUNCTION statement can and should be spread over several lines, to improve readability.

### 1.1.3 *Error Checking*

The interface language generates error messages automatically if the actual arguments (given by the user) do not match the formal arguments in the FUNCTION statement (too few, too many, or the wrong names), or if the actual arguments can not be coerced to the specifications given. For simple functions no further checks may be required. If the function requires that lengths be consistent or that data values fall in a specific range, or has any other substantive restrictions, the interface routine should check explicitly. If any requirements are not met, the interface routine should issue a meaningful error message. An error message in an interactive environment gives the user a chance to correct the expression and try again. There is no need to go to great lengths to produce a result from bad input data.

Error conditions can be checked by computing expressions

involving the arguments. An error message can be printed (and execution of the S expression terminated) by

FATAL(message)

Error checking usually involves attributes of the argument data structures, such as LENGTH, NROW, NCOL, and MISSING.

In the *fit* example, we require that the number of rows in the $x$ matrix and the lengths of $y$ and *wt* match. Failure of these requirements should generate an error and a message to the user through FATAL:

```
if(NROW(x)!=LENGTH(y))
    FATAL(Number of observations in x and y must match)
if(!MISSING(wt) & NROW(x)!=LENGTH(wt))
    FATAL(Weights must match number of observations)
```

The name of the function is added to the message automatically, so that, in the second case, the user would see:

Error in fit: Weights must match number of observations

### 1.1.4 *Dynamic and Static Data Structures*

Arguments to S functions are dynamically allocated, self-defining S data structures. Any results returned by a function must also be such structures. Function arguments may be returned as results. If a function is to return a result *other* than an argument, this must first be allocated as a dynamic structure:

STRUCTURE( $arg_1$, $arg_2$, ... )

where, as with the FUNCTION statement, $arg_1$, etc., are of the form:

name/ $spec_1$, $spec_2$, ... /

In the STRUCTURE statement, the specifications must define the actual length or similar description of the structure, as well as the type and mode. These actual parameters are given by a computational expression which may depend on attributes of other structures; these parameters need not be constants. In the *fit* function,

```
ncoef=NCOL(x)
if(int) ncoef=ncoef+1 #for the intercept
STRUCTURE(
        coef  /REAL,ncoef/
        resid /REAL,LENGTH(y)/
        )
```

allocates vector structures *coef* and *resid* based on attributes of the arguments *x* and *y*. The attributes, like the length of *coef*, are expressions computed at run time.

Allocated structures need not be used as results of the function. If an algorithm requires scratch space of variable size, this can also be more conveniently allocated through a STRUCTURE statement than through a static declaration.

When the allocated structure is a matrix, two parameters need to be provided, the number of rows and the number of columns:

STRUCTURE( xtrans/MATRIX,NCOL(x),NROW(x)/ )

The order of the parameters determines their interpretation. For time-series structures, three parameters are required: the start time, end time and number of values per year.

An alternative specification is frequently useful: when an allocated structure should be identical to an existing structure, the short form

LIKE(name)

may be used in place of any other specification. In the *fit* example, a better specification of the *resid* structure would be:

STRUCTURE( resid/LIKE(y)/ )

Not only is this more compact to write, but it ensures that *resid* inherits all the attributes of *y*; for example, if *y* is a time-series, the residuals will be also.

It is not necessary to make scratch copies of arguments to S functions. Arguments are already dynamic data structures and can be overwritten without harming user datasets.

Initial data values may be given in an additional parameter to the STRUCTURE specifications, or may be entered later.

Declarations for scalars, FORTRAN functions and other non-S data are identical to similar declarations in RATFOR or FORTRAN, except that these are STATIC data items in the interface language. The whole FORTRAN-style declaration must be included in a STATIC statement; e.g.,

STATIC( real xlim(2),mean; integer i,j )

which generates corresponding declarations in the FORTRAN code produced from the interface routine. Similarly, FORTRAN common block declarations should be included in a STATIC statement. The rules about what needs to be declared and how are otherwise identical to those of RATFOR and FORTRAN.

All S dynamic data structures are self-describing. This means that their type, mode, length, etc. are available at execution time by examining the structure. The interface language provides a number of attribute expressions to help in this process. The full set will be discussed in section 1.2. Some useful attributes are

> MISSING(x) #was argument supplied
> LENGTH(x)
> NROW(x), NCOL(x) #number of rows, columns
> TSTART(x),TEND(x),TNPER(x) #time-series parameters

Expressions to access individual data values of vectors, matrices, etc. may be specified as in the S language:

> x[i,j]
> y[i−1]

but within the interface language all subscripts must be positive integer scalars.

All other uses of the name of the data structure (for example, in a call to a subroutine) are interpreted as a reference to the first element of the vector of data values, e.g., *x* is equivalent to *x[1]*.

### 1.1.5 *Computations*

The interface language allows the full syntax of RATFOR to express computations. However, computations for S functions should primarily be carried out in algorithms called from the interface routine. This approach makes it likely that the algorithms will be useful more generally, and usually increases the efficiency of the resulting S function. Subroutines are called just as in RATFOR, with the added ability to reference S data structures and their attributes as illustrated above.

Suppose the *fit* function calls a subroutine *lsfit*:

> subroutine lsfit( x, nr, nc, y, coef, resid)
> real x(nr,nc), y(nr), coef(nc), resid(nr)
>
> ...

The interface routine calls this subroutine, using the arguments to the S function and their attributes to define the information needed by the subroutine:

```
call lsfit(x,NROW(x),NCOL(x),y,coef,resid)
```

When the algorithm *lsfit* has finished its computations, the data values in *coef* and *resid* are the desired coefficients and residuals. Now the S function *fit* wants to return a structure with components named *coef* and *resid*, which is done by the RETURN statement in the interface language:

```
RETURN(coef,resid)
```

The complete *fit* interface routine for a simple version of the function could look like:

```
FUNCTION fit(
        x/MATRIX/
        y/REAL/
        )
if(NROW(x)!=LENGTH(y))
        FATAL(Number of observations in x and y must match)
STRUCTURE(
        coef/REAL,NROW(x)/
        resid/LIKE(y)/
        )
call lsfit(x,NROW(x),NCOL(x),y,coef,resid)
RETURN(coef,resid)
END
```

### 1.1.6 *Creating New Functions*

At this point, enough of the interface language has been described to write an interface routine corresponding to a simple S function. It remains to take the interface routine, along with any other routines needed, and turn them into an executable S function. A set of utility commands exist to do this. They are executed at the UNIX shell level.

In order to develop a set of S functions, one must create a *chapter*, by using the utility

```
$ S CHAPTER
```

CHAPTER creates subdirectories in the current directory for the executable versions of functions and for documentation; it is used only once to make an S chapter in a given directory. A new function within the chapter is announced once by the command

> \$ S FUNCTION name name.i file1 file2 ...

with *name* the name of the function as called in S, and *name.i*, *file1*, etc. the names of files containing the interface routine and any other routines (algorithms) on which the function depends. The following notes about using FUNCTION should be kept in mind:

- The FUNCTION utility has the option "−r" for functions needing to read in the values of the S internal parameters for its computations (typically because it does printing or generates random numbers). In this form the function is declared

  > \$ S FUNCTION −r name name.i file1 file2 ...

- Similarly, if the function produces graphical output (section 1.5), or if it is a graphics device driver (section 3.6) the options "−g" or "−d", respectively, must be included.
- The interface routine must be written to a file with a name ending in ".i", as shown. Routines in RATFOR, FORTRAN, and C must be written to files with names ending in ".r", ".f", and ".C" respectively.† Routines in more than one language should not be included in the same file.
- The files named can also include object files or libraries (archives), so that your function can use algorithms of (FORTRAN-callable) routines developed for other purposes. However, the libraries of algorithms included with S itself are searched automatically and don't need to be mentioned.
- The first name given to the FUNCTION utility specifies the name by which the function will be called within S, and this must match the name in the FUNCTION statement of the interface routine.

We can illustrate these rules for the example of the *fit* function. Suppose the interface function is stored in file *fit.i* and the (RATFOR-based) source for the algorithm is on *lsfit.r*. Furthermore, suppose we want to use some routines already compiled into the archive file */usr/friendly/lib.a*. Then the FUNCTION utility is invoked by:

> \$ S FUNCTION fit fit.i lsfit.r /usr/friendly/lib.a

Once the function has been announced and the source code written for the interface routine and other routines, the function can

---

† C language routines that use S algorithm-language facilities should be on ".C" files. They will be processed by the *m4* macro processor. Standard C language routines that do not require this macro processing should be on ".c" files.

be created by

> $ S MAKE name

This initiates a number of steps to compile the interface routine and
other routines needed and to load the executable version of the function. As the name of the utility suggests, this is a version of the UNIX
system *make* program (Stuart I. Feldman, "Make - A Program for Maintaining Computer Programs", *Software—Practice and Experience*, Vol. 9,
pp 225-265, 1979). The FUNCTION utility creates rules, stored in a
file named *Smakefile*, by which the executable form of the function is
created. Because of special rules used in making S functions, however,
the S MAKE utility must be used rather than *make*.

Once the function has been made, it can be tested from within
S. The S function *chapter* is used to put the user's chapter on the
search list for functions. Specifically, if the chapter was created in the
same directory used to run S, the S command

> **> chapter**

can be used without arguments. Otherwise, the argument to *chapter*
must be the name of the directory in which the chapter was created:

> **> chapter("/usr/abc")**

(The UNIX command *pwd* gives the full pathname of the current directory.) Also, the function

> **> search(1)**

gives the list of chapters currently being searched by S, including the
standard S chapter.

After testing, changes may be made in the interface routine or
in other routines. Whenever changes are made, the executable version
of the function should be brought up to date by repeating

> $ S MAKE name

Ordinarily, the rules stored in *Smakefile* are sufficient to allow S MAKE
to recompile the function. However if the function is revised and
depends on a different set of files, the FUNCTION command must be
used again to update the rules. For example, if the *fit* function is
revised to allow input weights and now needs to call the additional
algorithm *lswfit*, stored on the file *lswfit.r*,

> $ S FUNCTION fit fit.i lsfit.r lswfit.r /usr/friendly/lib.a

There are several utilities to assist in creating documentation
for S functions. The utility

⁣$ S PROMPT file.i ...

constructs a skeleton documentation file (named *"file.d"*) for corresponding functions. These files contain embedded instructions preceded by the character *"~"*, that describe what should appear at that point in the documentation file. The documentation file should be edited to contain reasonable descriptions of the arguments, results and other significant properties of the function.

Once the documentation file has been prepared, it is entered into the chapter documentation by

⁣$ S NEWDOC file.d ...

After NEWDOC is used, the documentation file can be removed. Any user searching the chapter will now have access to the documentation via *help*.

To make changes in the documentation once it has been installed in the chapter, use

⁣$ S EDITDOC −f file

which will recreate *"file.d"*. This file can be edited, and then re-installed by NEWDOC.

Function documentation can be printed on the user's terminal by

⁣$ S PRINTDOC −f [name ...]

All documentation for the chapter is printed unless the list of names appears.

**1.1.7** *Hints; Debugging*

The fundamental maxim remains: KEEP IT SIMPLE. Try to start with an uncomplicated interface routine and a straightforward set of underlying algorithms. Let the rest of the S language do the work of reading in information, printing results, and organizing data. In particular, it is not a good idea at this stage to print anything other than the error messages shown above, and still less desirable to try to read data values from a file.

S has facilities, however, for printing debugging information, either from an interface routine or from an underlying RATFOR algorithm. The statement

DEBUG

in either case invokes, during the execution of the S function, an interactive debugging facility. The user responds by typing the name

of an S structure that occurs in the interface routine. The current contents of the structure will be typed on the user's terminal, in the format used by the S function *dput*. (This is not as pretty as S output, but
is more general and is adequate to see the current values in the structure.) To exit from DEBUG, type an empty line. You can interrupt the
printing of a dataset; DEBUG will return to prompt for more input.

Notice that the name typed to DEBUG is the *keyname*, the name
which appears in the FUNCTION or STRUCTURE statement, not the
name of the actual data set supplied. To see a table of the available
argument names, type "?" to DEBUG. The response will be a table
with keynames, datanames, and an indication of whether the
corresponding argument was MISSING in the actual call.

In the *fit* function, suppose the DEBUG statement is inserted
just in front of the call to *lsfit*. Then a session in the use of the function might be

> > **fit(pred[,1:2],response)**
> Debug called from fit:
> Debug: **?**

```
Keyname         Dataname        Missing?
x                               F
y               response        F
coef                            F
resid                           F
```

> Debug: **x  y**
```
x:
  ( "" S 21 2
    ( "Dim" I 2   16   2 )
    ( "Data" R 32    83   88.5   88.2   89.5   96.2   98.1     99
        100   101.2   104.6   108.4   110.8   112.6
      114.2   115.7   116.9   234.289   259.426   258.054   284.599
      328.975   346.999   365.385   363.112   397.469
      419.18   442.769   444.546   482.704   502.601   518.173
      554.894 )
  )
y:
  ( "response" S 0 1
    ( "" R 16   60.323   61.122   60.171   61.187   63.221   63.639
      64.989   63.761   66.019   67.857   68.169
      66.513   68.655   69.564   69.331   70.551 )
  )
```

Even when invoked from the RATFOR routine, the interactive DEBUG
feature dumps only the S data structures, as known in the interface

routine. Since these are generally the data passed down to algorithms, however, one can monitor the computations by knowing how the algorithms were called.

In either interface routines or RATFOR routines, it is also possible to use DEBUG in an non-interactive way to print out single datasets. To print an S data structure in an interface routine, say

DEBUG(name)

with *name* again the keyname of the S structure. For non-S data, such as variables in a RATFOR routine,

DEBUG(data,mode,length)

where *data* is some expression, *mode* is one of the data modes (REAL, INT, LGL, CHAR) and *length* is the number of values to print out.

The use of DEBUG, plus careful planning and a gradual approach to introducing complexities should make the debugging of new functions relatively straightforward. Interactive debuggers included in the UNIX environment (e.g., *sdb* or *adb*) cannot be used directly in debugging an S function. Even if they could, the symbolic names known to them would mostly be those generated as output from the interface language, and these are not readable by humans. However, one useful application of UNIX debuggers occurs when one wants to debug an underlying algorithm, rather than the S function. Then the following trick allows the debugger to be used: suppose we want to run *sdb* on the function *fit*. Write, say on the file *dbfit.c*, in the directory containing our chapter, the trivial C program:

```
main() { system("sdb x/fit"); }
```

Now create the file *x/dbfit*, containing the executable version of this; i.e.,

**$ cc −o x/dbfit dbfit.c**

To use the *sdb* version from S, call *dbfit* with the same argument list that would have been used with *fit*. Note again, however, that the debugging will not likely be useful in interface language routines; the breakpoints should be put in algorithm code (and UNIX debuggers tend to do better with C than with FORTRAN).

**1.1.8** *Strange Happenings; Warts*

Because the interface and algorithm language compilers are constructed from several separate programs, the languages inherit a number of idiosyncrasies of these tools. In particular, the algorithm language uses the *m4* macro processor and the *ratfor* FORTRAN pre-processor. The interface language uses both of these programs, along with a specialized processor of its own.

- *m4* doesn't recognize quotes as delimiters or anything special, thus

        macro("this is a string, containing a comma")

    is interpreted as a macro with two arguments (see section 2.2.1).
- argument lists to *m4* macros must appear immediately after the macro name; there can be no intervening white space; thus

        macro(arg)
        macro (arg)

    are very different in effect. The first gives one argument to the macro; the second invokes the macro with no arguments.
- The characters grave ( ` ) and apostrophe ( ´ ) are used as quotes for *m4*, and should not be used for other purposes.
- Because of limitations of the interface language, arguments to the RETURN statement should not have trailing blanks.

## 1.2 Data Structures

This section discusses the data structures which can be defined in the interface language.

### 1.2.1 *Data Values and Attributes*

Good use of the interface language should leave most of the actual computations on data to underlying algorithms. Occasionally, it may be more convenient to do some computation in the interface routine, particularly when working out default actions.

Individual data elements in S data structures are specified by using square-bracket notation, as in S:

    x[i,j]=1
    amax1(y[i],0.)

Since the references must eventually be translated into FORTRAN, only single values can be worked with. When calling subroutines, column $j$ of the matrix $x$ is passed as $x[1,j]$ and row $i$ as $x[i,1]$ (of course, individual data items in a row are spaced NCOL(x) apart).

Since all datasets are self-describing, they have *attributes*; i.e., values which define the structure in S. Vectors and vector structures (including matrices, arrays and time-series) will have the attributes:

> MODE(x)
> LENGTH(x)

In addition, matrices will have NROW(x) and NCOL(x) for the number of rows and columns. Time-series will have the start, end, and number of observations per time period as the (real valued) attributes TSTART(x), TEND(x) and TNPER(x). The function *diff1* defined below, shows the use of time-series attributes and also an example of computations which are actually simpler in the interface language, since they largely involve manipulation of attributes.

```
FUNCTION diff1(
        x         /TS/
        lag       /INT,1,1/
        )

STATIC( real start; integer i )
if(lag>=LENGTH(x) | lag<1)
        FATAL(Number of lags not between 1 and length of data)
start=TSTART(x)+float(lag)/TNPER(x)
STRUCTURE(y/TS,start,TEND(x),TNPER(x)/)

for(i=1; i<=LENGTH(x)-lag; i=i+1)
        y[i]=x[i+lag]-x[i]

RETURN(y)
END
```

### 1.2.2 *Character Data and Character Attributes*

Computations on real, integer, and logical data values map down to FORTRAN with little change. Computations with character string data, however, require some additional discussion, since FORTRAN has only limited facilities to deal with character data.

A character vector in S is a vector of integer *pointers* to character strings. The text in each character string is terminated by the special character "\0", which can be referred to symbolically as "EOS". In thinking about character data, it is important to distinguish the *pointers* to the text from the text itself. The character vector in S is actually a vector of pointers (integers, actually, but that doesn't matter); it is these pointers that are rearranged, extracted for subsets and manipulated in various ways. When the actual text corresponding to the pointer is to be used, the attribute TEXT generates a reference to that. For example, if *labels* is a structure of mode CHARACTER, then labels[i]

is the (pointer to) the i-th string, and

　　　　TEXT(labels[i])

is the actual text.

　　　　S functions often have character arguments allowing users to supply titles or labels for printing or plotting. Default values can be generated for these arguments. The simplest case is that of a single character string, such as the title for a plot. When this is an optional argument, a default pointer to a chosen string can be allocated by the attribute STRING:

　　　　FUNCTION qqnorm(

　　　　　　...

　　　　　　xlab/CHAR,1,STRING(Normal Quantiles)/

　　　　　　...

The argument *xlab* defaults to a character vector of length 1, pointing to the string "Normal Quantiles".

　　　　When a vector of labels is needed, some other computations will be required. A general technique is to encode the labels from various data values. This generally requires use of the ENCODE statements in the algorithm language, to be described in section 2.2.1. These calculations are best done in an algorithm, not in the interface routines. A simpler case arises when the user supplies a character argument to be matched against one of several possible values. This, for example, is a much nicer way to have users choose one of several methods than asking the user to remember a numerical code for each method. The vector of possible strings is created by the CTABLE statement in the interface language. To declare and initialize a vector of character strings, use the statement:

　　　　CTABLE(table,entry$_1$, ..., entry$_k$ )

This initializes *table* as a (STATIC) vector of *k* pointers to character strings. The first element of *table* points to a character string containing *entry$_1$*, etc. For example, in the S function *hclust*, the user can select from three possible methods [Becker and Chambers (1984), page 322]. In the corresponding interface routine, a table against which to match the user's choice is created by:

　　　　CTABLE(meths,average,connected,compact)

creating vector *meths* of length 3 containing pointers to the strings "average", "connected" and "compact". To look up argument values in such a table, use the *match* function. This applies the partial match strategy used to match S argument names (see section 1.3.1).

    FUNCTION clust(

      ...

    method/CHAR,1,STRING(compact)/

      ...

    CTABLE(meths,average,connected,compact)
    i=match(method,meths)
    if(i<0)FATAL(bad value for method)

The value returned by *match* will be $-1$ if no successful match was possible, and otherwise will identify which value in the table matched. Note that the arguments to *match* are the pointer to the actual value and the vector of pointers defining the table. The CTABLE statement marks the end of the table by a pointer having the special value USED, so that the length of the vector is not needed by *match*. It is possible to construct such tables by any method desired, although the CTABLE statement is typically the simplest and most convenient.

    Certain character pointers are defined as attributes in the interface language. For example, FNAME is a pointer to the name of the S function. For each data structure in the interface routine, KEYNAME(x) is a pointer to its structure keyname and DATANAME(x) is a pointer to the name of the actual argument, if it had one. Thus, in the use of DEBUG on page 15, the keyname of *y* was "y" but the dataname was "response", from the actual argument. Again, these attributes are mainly useful in printed output, plotting and occasionally in error messages. For example,

    FUNCTION qqnorm( y/REAL/

      ...

    ylab/CHAR,1,DATANAME(y)/

      ...

uses as default value for the argument *ylab* the name of the actual dataset given to *qqnorm* for argument *y*. (If the actual argument is a general expression, it will not have a dataname. In this case, DATANAME(y) points to a null string containing only the EOS character.)

### 1.2.3 *Missing Values*

S data structures can contain missing values (NAs). By default, however, any argument to an interface routine will be checked for missing values, and an error generated if any are found. This reflects our feeling that there are no reasonable default strategies for dealing with missing values across the breadth of data analytic techniques. Functions that feel they know how to deal with missing values bear the onus of specifically allowing them. In the declaration of an argument or other structure, this can be done by the specifier, NAOK, indicating that missing values are allowed.

One possible way of handling NAs is to pass them on in corresponding values of the result of the function. Consider the function *diff*, for example. It is reasonable to allow missing values in the argument, and to set a value to NA in the result if either of the data values subtracted to get the difference are missing. Two further expressions in the interface language allow this: NA(expr) returns TRUE or FALSE according to whether the argument value is missing, while NASET(item) sets its argument to missing. (NA is a macro and hence should *not* be declared logical.) Missing values should not be assigned or otherwise operated upon in the interface routine: they should only be handled by NA or NASET. In the example of *diff*, the description of the argument and the calculation change as follows:

```
FUNCTION diff( x/TS,NAOK/, ...
     ...
for(i=1; i<=LENGTH(x)−lag; i=i+1)
        if(NA(x[i]) | NA(x[i+lag]) ) NASET(y[i])
        else y[i] = x[i+lag]−x[i]
     ...
```

If either *x[i]* or *x[i+lag]* is missing, so is *y[i]*; otherwise we compute the difference.

### 1.2.4 *Structures and Components*

S has general hierarchical structures, whose components are extracted in the S language by using the "$" operator. Techniques are available for operating on hierarchical structures in the interface language as well. The type STR in the description list corresponds to a general hierarchical structure. One can then refer to the component (if any) of a structure *z* having the name *x*, say, by:

x/FROM(z),.../

This allows reference in the interface language to what would be denoted by z$x in S. For example, consider the function *regsum* [Becker and Chambers (1984), page 416]. This function takes the hierarchical structure produced from a regression and uses various components of the structure to derive additional summaries of the regression. The regression structure is the argument to *regsum*. In the interface routine, it is necessary, for example, to extract the components *coef* and *resid* from the regression structure. This could be done as follows:

```
FUNCTION regsum( z/STR/ )
    ...
STRUCTURE(
    coef/MATRIX,FROM(z)/
    resid/MATRIX,FROM(z)/
    )
    ...
```

The argument z is coerced to be a hierarchical structure. Then the structure *coef* is defined by looking for the component called *coef*, and coercing this to mode MATRIX. If there is no component name matching *coef*, an error is generated. It is also possible to make the search optional, just as in the description of arguments to the function. For example,

```
STRUCTURE( int/FROM(z),LGL,1,TRUE/ )
```

looks for *int* as a component of z, but allocates a single logical value (TRUE) if there is no component matching *int*.

Notice that the FROM component is actually part of the original structure; changes in the component change the structure. If what is wanted is actually a separate copy of the component, use the COPY specifier:

```
STRUCTURE(x/FROM(z),REAL,COPY/)
```

Another variation on FROM allows us to rename the component. For example, if we wanted to look for the component *coef* in two different structures, we can't call both of the derived structures *coef*. A second argument to the FROM specifier says what name to look for, regardless of the name of the structure being generated:

```
STRUCTURE(coef2/FROM(z2,coef/)
```

makes *coef2* refer to the *coef* component in the second structure.

The process of extracting components from a structure is very much like that of looking for arguments. It is possible to use the standard argument processing facilities to examine the components of a structure. The chief application of this technique is for functions which can take either a list of arguments or a structure whose components are treated as arguments. Section 1.3.4 will discuss the details of the technique.

### 1.2.5 *Modes Computed at Execution*

*(This is a specialized, advanced topic. Skip it unless you need it.)* In the discussion and examples so far, the mode of the data in a structure was constant (i.e., known at the time the interface routine was compiled). Interface routines should be organized so that this is the case, whenever possible. The interface language automatically generates references to the correct data mode in response to the use of the name of the structure or an expression like *x[i,j]*. If the mode is not known until execution, the writer of the interface must take special steps to ensure that a check of the actual mode is made and that correct run-time code is generated for each case. The interface language does not do this automatically.

There are, however, functions that do allow different modes at execution. One example is the *encode* function. This S function returns a character vector produced by encoding any non-character arguments and then concatenating corresponding data elements in the arguments; e.g.,

```
> encode("Value",1:3)
```

```
"Value 1"   "Value 2"   "Value 3"
```

The arguments to *encode* can be of any mode, and one obviously wants to encode each argument correctly for each mode. Here is a simplified version of the function, which takes two arguments.

```
FUNCTION myencode(  x/ANY/, y/ANY/ )

STATIC( integer i,ii,nmax; POINTER istrng )
nmax=max0(LENGTH(x),LENGTH(y))
STRUCTURE( c/CHAR,nmax/ )

do i=1,nmax {
      ii = mod(i-1,LENGTH(x))+1
      SWITCH MODE(x) {
      CASE REAL: ENCODE( R(x[ii]) )
      CASE INT: ENCODE( I(x[ii]) )
      CASE LGL: ENCODE( L(x[ii]) )
```

```
CASE CHAR: ENCODE( C(TEXT(x[ii])) )
        }

ii = mod(i-1,LENGTH(y))+1
SWITCH MODE(y) {
CASE REAL: ENCODE( R(y[ii]) )
CASE INT: ENCODE( I(y[ii]) )
CASE LGL: ENCODE( L(y[ii]) )
CASE CHAR: ENCODE( C(TEXT(y[ii])) )
        }
c[i] = istrng(BUFFER,BUFPOS)
CLEAR
}

RETURN(c)
END
```

Several new techniques appear in this example, some of which will be treated in more detail later. The arguments are declared mode ANY (and by default are vector structures). They will not be coerced to any specific mode, as long as they can be treated as vectors. The interface routine will produce a result, $c$, which is as long as the longer of the arguments.

The *do* loop encodes individual data elements from $x$ and $y$ and creates a character string pointer to the concatenated result. This pointer becomes the corresponding data value in the vector $c$.

Data values in $x$ and $y$ are reused cyclically, if the arguments are of different length (this is the purpose of the code to compute *ii*). The encoding of different elements is done by means of algorithm language ENCODE expressions, to be discussed in section 2.2.1. The point to note here is that the computations for different modes are done by means of the SWITCH MODE expression:

SWITCH MODE(x) {

CASE REAL: #the computations for the case that x is REAL

   ...

The construction is similar to the *switch* statement in RATFOR, but is specially processed by the interface language so that, in the code which follows CASE REAL, all references to $x$ treat data values as real, and similarly for the other modes. This construction is the essential technique for handling modes which vary at execution.

A corresponding situation arises when a data structure is to be allocated with a mode that is not known until execution time. Some such cases can be handled by LIKE, but there is also a mechanism to allocate a structure with a mode calculated during execution; namely,

STRUCTURE( y/MODECALC(expression), .../)

The argument to MODECALC can be any computed (integer) expression; typically, it is MODE(x), where $x$ is some other data structure. The following example, however, illustrates another situation. In the function *pmax*, we want to find the maximum data value of all the arguments. Suppose, again, we restrict the function to two arguments. A simple way to write the function would be as follows:

```
FUNCTION mypmax( x/REAL/, y/REAL/ )

STATIC( integer i,nmax,ix,iy )
nmax=max0(LENGTH(x),LENGTH(y))
STRUCTURE(z/REAL,nmax/)

do i=1,nmax {
        ix  =   mod(i-1,LENGTH(x))+1
        iy = mod(i-1,LENGTH(y))+1
        z[i] = amax1( x[ix],y[iy] )
        }
RETURN(z)
END
```

As with *encode*, the data elements of $x$ and $y$ are used cyclically. Notice that *mypmax* will return a result of mode REAL even if both its arguments were INT. There is nothing seriously wrong with this and user functions might well be better to keep things simple. However, for large integers, there may be a loss of precision on converting to real values. For this and for aesthetic reasons, one might prefer to return integer values when the arguments were integer. This can be done with MODECALC.

The idea is that if both arguments are either logical or integer, the result will be integer, but if one or more arguments are real, so is the result. (To simplify, logical arguments will be coerced to integer; there is no loss of information.) We take advantage of the fact that the built-in modes in S actually have numerical values (try printing the S built-in datasets REAL, INT, LGL, etc.). The numerical values are chosen to reflect a hierarchy useful in coercing:

LGL < INT < REAL < CHAR

Thus, the mode of the results is just the maximum of the modes of the arguments. Since logicals are being coerced to integers, the mode will be at least as large as INT. Thus, the computed mode is

mode=max0(INT,MODE(x),MODE(y))

When *mode* is computed, it implies not only the mode in which the result should be allocated, but also the mode to which $x$ and $y$ should be coerced. This introduces another new idea: coercing data structures after they have been found as arguments or allocated. This is done by the COERCE statement:

COERCE( name/type,mode/ )

where *type* and *mode* can be any of the types and modes introduced previously.

It would be possible to use MODECALC as a mode in the coerce statements. However, we will have to use SWITCH MODE to handle the different modes in any event. These ideas lead to the following version of *mypmax*.

```
FUNCTION mypmax( x/ANY/, y/ANY/ )

STATIC( integer i,nmax,mode,ix,iy )
nmax=max0(LENGTH(x),LENGTH(y))
mode=max0(INT,MODE(x),MODE(y))
STRUCTURE(z/MODECALC(mode),nmax/)

SWITCH MODE(z) {
CASE REAL: COERCE(x/REAL/); COERCE(y/REAL/)
      do i=1,nmax {
             ix=mod(i-1,LENGTH(x))+1
             iy=mod(i-1,LENGTH(y))+1
             z[i]=amax1( x[ix],y[iy] )
             }
CASE INT: COERCE(x/INT/); COERCE(y/INT/)
      do i=1,nmax {
             ix=mod(i-1,LENGTH(x))+1
             iy=mod(i-1,LENGTH(y))+1
             z[i]=max0( x[ix],y[iy] )
             }
      }
RETURN(z)
END
```

Modes computed at run time must be used carefully, and should be avoided if possible. The basic rule is that one can only compute with data items in the interface language when their mode is known. Thus, if the mode of a structure is not known at compile time, a SWITCH MODE expression must be used, with separate and essentially duplicate code for each mode of interest. Where other structures are involved, it is usually necessary to coerce these to the corresponding modes. The interface language's knowledge of modes is determined strictly by position in the interface routine. After encountering a statement which says that the mode of $x$ is REAL, each succeeding statement referring to data values of $x$ treats those as REAL. If there were a jump into the middle of this code during execution, when $x$ was not actually REAL, chaos would result.

There are a small number of special algorithms accessible to interface routines which treat data structures of arbitrary mode. The technique is to pass these routines the attribute

VALUE(x)

which is a pointer to the data values. Along with the mode, this allows the algorithms to treat the data correctly. An example of such an algorithm is

call pcopy(from,to,length,mode)

which copies data between two data structures, given the value pointers *from* and *to*. New algorithms of this type can be written as well. This approach may turn out to be a good solution to handling data of arbitrary mode. Without trying to go into detail, the following code for the *pcopy* algorithm illustrates the kind of calculation possible:

```
subroutine pcopy(from,to,length,mode)
POINTER from,to;  integer length,mode
INCLUDE(stack)
integer i

if(length<0) ERROR(no. of items < 0)
for(i=0; i<length; i=i+1) {
    if(NAVALUE(from+i,mode))NASET(is(to+i))
    else switch(mode) {
        case REAL: rs(to+i)=rs(from+i)
        case INT,CHAR: is(to+i)=is(from+i)
        case LGL: ls(to+i)=ls(from+i)
        default: ERROR(bad mode in copy)
            }
    }
return
end
```

(See section 2.2.4 for the *stack* facilities.)

## 1.3 Function Arguments

This section deals in detail with the processing of arguments to S functions. It explains how argument matching occurs, how the names of arguments can be abbreviated by the user, how computations can be carried out in the middle of argument processing, how functions can accept arbitrary numbers of arguments, and how arguments can be extracted from data structures.

### 1.3.1 *Arguments in the FUNCTION Statement*

As we have seen in the previous sections, the list of arguments in the FUNCTION statement is of the general form:

( name$_1$ /description list$_1$/, ... name$_k$/ ... /)

When the interface routine for the function is invoked, the *actual arguments* (given in the user's call) are scanned to find the argument corresponding to each of the *formal arguments* (in the definition of the function) in turn. If an actual argument of the *name=value* form matches *name$_i$* exactly, or by an unambiguous partial match (see below), this actual argument is used for the formal argument *name$_i$*. Otherwise, the first unnamed actual argument not previously matched is used. If no actual argument matches, the attribute MISSING(name$_i$) is set to TRUE and the default action is taken. If a default structure was defined, this is allocated. If OPTIONAL was specified, but no default structure, the structure is allocated of type NULL (presumably something will be done later to check for the argument being missing). If neither default form was specified, a FATAL error is generated.

The partial match procedure works as follows. The interface routine has a table of the names of all the formal arguments appearing anywhere in the function. To match a specific formal argument, the actual arguments are scanned to find either an exact match, or an actual argument whose name matches a non-empty leading part of the name of this formal argument (and of *no other* formal argument). For example, suppose the formal arguments of a function have names "x", "y", "intercept" and "iter". Consider the following actual arguments:

    i=1
    int=1
    interdept=1

The first fails to match any argument, since its name is a leading part of both "int" and "iter". The second successfully matches "intercept". The third fails to match any argument (a partial match with spelling correction might help.) The idea of the partial match is to allow users to employ short abbreviations, but permit the formal names of the arguments to be informative. That is why, for example, many S functions use the plural form of argument names, for example, "labels"; the user can give the argument as "labels", "label", "lab", or even "l", as long as the argument is uniquely identified.

When the search and coerce are completed for all the arguments in the argument list, the operation ENDARGS is invoked automatically to check that all the actual arguments have been matched to some formal argument (to detect extra arguments or misspelled argument names).

**1.3.2** *Interrupting and Resuming Argument Processing*

It may be useful to make the argument processing depend on some computed tests; for example, on whether or not some prior argument was supplied. The ENDARGS check may be suppressed by giving "&" as the last argument:

```
FUNCTION abc(
    x  /MATRIX/
    wt  /REAL,OPTIONAL/
    &
    )
```

This special symbol does not match any actual argument, but simply indicates that the remaining arguments (if any) should be left for further processing. By including "&" as the last argument in the FUNCTION statement, the interface routine can now do any additional computations. Argument processing is resumed by the statement

$$\text{ARG( name}_1/ \ ... \ /, \ ..., \ \text{name}_k/ \ ... \ / \ )$$

Searching and coercing in the ARG statement are identical to the same process in the FUNCTION statement.

For example, suppose function *abc* takes an optional argument *wt*. If *wt* is present, a further optional argument, *wtmin*, may be supplied.

```
FUNCTION abc(
    x       /MATRIX/
    wt      /REAL,OPTIONAL/
    &
    )

if(!MISSING(wt)) {
    ARG( wtmin/REAL,1,0./, & )
    ...
    }
ENDARGS
```

Notice that the ARG statement, like the FUNCTION statement, will finish by performing the ENDARGS operation unless ARG also gets a final argument "&". In the example, to check, consistently, for the end of the argument list, the interface routine invokes ENDARGS explicitly.

It is possible to suppress positional matching of arguments and require arguments to be given only in the *name=value* form, simply by following the name by "=" in the FUNCTION statement:

FUNCTION display( title=/CHAR,OPTIONAL/, ...

An actual call of the form

display("Hello", ...)

would *not* match "Hello" to the formal argument *title*. Arguments specified in the interface routine in the *name=* form are not matched partially: the actual name must be given exactly (this is useful when a function can take arbitrarily many arguments; see below).

### 1.3.3 *Arbitrarily Many Arguments*

S functions like *c, print,* and *save* take a list of arbitrarily many arguments and perform some computations on each of them. In the interface language, this requires a loop over all actual arguments, rather than a process of matching each actual argument to a different formal argument. The loop is defined by the matching pair of statements:

ALLARG( name )

...

NEXTARG

The effect is as follows. ALLARG looks for the next available argument (excluding those matched by previously executed FUNCTION or ARG statements). If there is no such argument, control breaks out to the statement after the matching NEXTARG statement. Otherwise, *name* specifies an S structure which is made to correspond to the matching actual argument. The statements down to the matching NEXTARG are now executed and control returns to the ALLARG statement.

This argument matching process differs in several ways from the FUNCTION or ARG statement. It is non-destructive: another subsequent ALLARG loop will match the same arguments in the same way. (However, setting NAME(x)=USED will mark the argument so that it will not be re-processed.) It ignores the names, if any, given to the actual arguments. It does not coerce the arguments to any specific type or mode.

The non-destructive loop is useful in that one often needs to go over the arguments twice, the first time to determine lengths and/or modes, the second time to allocate and process data.

Since ALLARG does not coerce the structure to a type or mode, neither special attributes (NROW, etc.) nor actual data values will be available automatically for the matched argument. The COERCE statement should be used in the loop to coerce the argument.

```
FUNCTION mixem( & )

...

ALLARG(x)
      COERCE(x/MATRIX/ )

      ...

NEXTARG
```

The description list in the COERCE statement is the same as in the ARG statement, except that none of the default description of the structure can be supplied.

Two situations arise with respect to argument names in the use of ALLARG. The simplest is that the argument names are irrelevant to this S function, and should not be supplied. In this case, the interface routine should not match arguments in the *name=value* form. The option NOKEY to ALLARG suppresses match of any named arguments:

```
ALLARG(x,NOKEY)
```

Named arguments in functions with arbitrarily many arguments are usually special arguments, matched outside the ALLARG loop. The NOKEY option prevents matching them in the loop and therefore allows checking for misspelling of the name, using ENDARGS:

```
FUNCTION display( title=/CHAR,OPTIONAL/, ..., &)

...

ALLARG(z,NOKEY)
      NAME(z) = USED

      ...

NEXTARG

# catch misspelled title or unrecognized keyword arg
ENDARGS
```

Notice that here we recognized *title* only in the *name=value* form. This is usually convenient, since the user can supply *title* anywhere in the argument list, with no danger of confusing it with the other, ordinary arguments.

A different situation occurs when the function uses the name given to the actual argument in computations. For example, the *save* function uses any argument names as the name for the corresponding dataset on the save directory. In this case, one allows named arguments and picks up the name, if it is not NULL, in the ALLARG loop:

```
FUNCTION save( pos=/INT,1,2/, &)

STATIC( POINTER pname )
ALLARG(x)
     if(NAME(x)!=NULL) pname=NAME(x)
     else pname=DATANAME(x)

     ...
NEXTARG
```

This form is suitable if the S function wants to take a name from the dataset, but wants to allow the user the option to override.

### 1.3.4 *Treating Structures Like Argument Lists*

The list of arguments to a function forms a hierarchical data structure, with the individual arguments being the components of the structure. Sometimes a process similar to argument matching is useful for the components of other structures as well. For example, a structure produced as output from one function may later be given to another function as an argument. In this case, the structure is *one* argument; all the results of the previous function need to be extracted from this structure. (Contrast the use of CHAIN (section 1.4.2), which passes individual results from the first function as individual arguments to the second.)

The most straightforward way to extract components from a structure is the FROM expression in a STRUCTURE statement (as discussed in section 1.2.4):

```
STRUCTURE( Dim/FROM(x),INT,OPTIONAL/ )
```

This looks in the structure *x* for a component whose name matches "Dim". The component, if found, is coerced according to the description list, just as it would have been in a FUNCTION or ARG expression (OPTIONAL and default descriptions may be used as well.) The attribute MISSING will be set according to whether the component is found.

For example, the function *bxp* takes as an argument the single structure *z*, which always has components named *stats*, *conf*, and *n*, and which may contain other components, such as *names*, as well. One way to implement this argument situation would be:

FUNCTION mybxp(z/STR/, &)

STRUCTURE(
        stats  /FROM(z),MATRIX/
        conf  /FROM(z),REAL/
        n      /FROM(z),INT/
        names /FROM(z),CHAR,OPTIONAL/
        )

If we need to pick up the same component from more than one source, FROM can be given a second argument, which is the name to look for in the structure:

STRUCTURE(
        adim/FROM(a,Dim),INT/
        bdim/FROM(b,Dim),INT/
        )

This statement generates structures *adim* and *bdim* corresponding to the component named "Dim" in *a* and *b* respectively. Section 1.2.4 gives more details on the FROM specifier.

The repeated use of FROM may be verbose if many arguments are to be extracted from one structure. The expression

ARGSTR(z)

shifts the entire argument-searching process to the components of *z*, rather than the original argument list. Subsequent ARG statements will always refer to the structure *z* until the next ARGSTR statement. In particular,

ARGSTR

returns argument searching to the original argument list.

There are several differences in the effects of FROM and ARGSTR. The argument search mechanism in the ARG statement ensures that each actual argument is matched only once, by setting the name field of the actual argument to the special value USED, so that it will not match anything on subsequent searches. In contrast, the FROM expression just searches for the required component.

After ARGSTR, names will be matched by the partial match strategy against the list of *all* argument names (not just those which were mentioned as possible components of z). The FROM expression does no partial matching; component names must match exactly.

FROM is simpler for most applications; ARGSTR is useful when the structure is really standing in for a (potentially complex) argument list as was the case with *mybxp* above.

## 1.4 Function Results and Related Statements

The RETURN statement in an interface routine is the mechanism by which the S function returns the data structure which is its result. As was the case with the FUNCTION statement, there are several ways in which RETURN can be used: it can control the name of resulting components, deal with unused arguments, handle graphical parameters, or CHAIN to another function. A related statement, INSERT, allows the construction of hierarchical data structures.

### 1.4.1 *The RETURN Statement*

The general form of the RETURN statement is one of:

RETURN( $comp_1$, ..., $comp_k$ )        # or
RETURN( $comp_1$, ..., $comp_k$, & )

The effect of the statement is to add the structures $comp_i$ as components of the data structure returned by the function. If "&" is included, execution of the interface routine then continues. Otherwise, the function returns to the S executive. The components $comp_i$ can be of the forms:

$structure_i$
$name_i$ = $structure_i$
= $structure_i$

where $structure_i$ is the name of an existing S data structure in the interface routine, and $name_i$, if present, is any legal name. In all cases, $structure_i$ is made a component of the result of the function. If no name is specified, the component name is $structure_i$ itself; in the second and third forms the given $name_i$ or an empty name is used as the component name. It is irrelevant whether $name_i$ is the name of an existing structure or not; it is only used for naming the component of the result generated.

As an example of the usefulness of renaming arguments, suppose that $x$ is an argument to a function, but that we wish to return a structure with a component called "x", containing different values from those in $x$; say, values in a vector *plotx*. This can be done, without re-allocating and copying, by:

RETURN(x=plotx, ... )

This renaming technique is also useful in functions that CHAIN to other functions (1.4.2).

In addition to the general forms above, RETURN takes two special arguments:

FILTER

PAR

The special component PAR returns graphical parameter settings (see section 1.5.2). The component FILTER returns as results all *unmatched* arguments to the S function. It is used in conjunction with "&" in the FUNCTION statement and the CHAIN statement to pass along unused arguments as results.

Here are some details of the way RETURN works. Structures are associated with components of the result through the dynamic structure pointed to by *structure$_i$* at the execution of the RETURN statement. The data values are not copied to a temporary structure; rather, the pointers to the structure are inserted into the result. This is usually irrelevant in practice, except for the implication that it is unwise to change the contents or to reallocate a structure appearing in a previous RETURN statement.

Only dynamically allocated structures may be returned, not scalars or arrays appearing in STATIC declarations.

If only one component is returned, this component becomes the entire result of the function, as opposed to returning a structure with a single component. Again, this is not usually an important distinction in practice.

One further note: if you write an S function whose primary purpose is to produce printed output (perhaps using the facility described in section 2.2.1), it is possible to have the function return a value that is not automatically printed. This is how the function *regprt* works, for example. This is done by executing the interface language statement

NOPRINT

somewhere prior to the RETURN statement. Since NOPRINT is an executable statement, it can also be done conditionally.

**1.4.2** *CHAIN: Invoking Another Function*

In a number of circumstances, a new S function is an extension of, or wants to use an existing function. Often, this can be expressed in the sense that the new function would like to finish by invoking the existing function. For example, the probability-plotting functions *qqplot* and *qqnorm* prepare two sets of co-ordinates for a scatter plot and chain to the scatter plot function *plot* to do the graphics. In the interface language, this is done by the CHAIN statement:

CHAIN(name)  #or
CHAIN(name, structure$_1$, ..., structure$_k$ )

The first argument, *name*, is the name of an S function. The remaining arguments, if any, are interpreted exactly as arguments to the RETURN statement. The effect of CHAIN is identical to that of RETURN, with the additional side effect that on return to the S executive, the function *name* will then be called. The components of the result of the current function will become the arguments (given in *name*=*value* form) to function *name*.

Notice that there is a difference between CHAIN-ing to a function and calling the function with the result of the previous function. For example, suppose the interface routine for function *abc* has the statement:

CHAIN(def,x,y,wt)

The effect is to call function *def* with three arguments, named *x*, *y* and *wt*. If, on the other hand, *abc* had the statement

RETURN(x,y,wt)

and was used in an S expression of the form:

def(abc( ... ))

then *def* is called with one unnamed argument, with three components. Either may be desirable, but the effect is different. See the INSERT statement (1.4.3) for simulating the second form.

The use of CHAIN is one reason for having the capability to rename the results of an S function (1.4.1). The names in the CHAIN statement may be chosen to be those required by the function to be invoked next, or may be empty to generate positional arguments.

### 1.4.3 *INSERT: Building Structures*

Corresponding to the ability to extract components from a structure (section 1.3.4), it is possible to insert structures as components of other structures.

INSERT(structure, component$_1$, ... )

The syntax of *component$_i$* is the same as for the RETURN statement. Each structure specified in the component arguments is inserted into *structure* as a component. As with the RETURN statement, the component will have the name specified in the *name=component* form if this form is used. The difference between RETURN and INSERT is that null names may not be used with INSERT.

## 1.5 Graphics Functions

S functions to do plotting are slightly different from other functions. They communicate, through graphical algorithms (chapter 3), with the device driver functions. They frequently take graphical parameters as optional arguments. Also, they often recognize a special argument structure which is useful for specifying plotting positions.

### 1.5.1 *Declaring a Graphics Function*

When a graphics function is first declared, the FUNCTION utility must be given the special option "−g":

$ S FUNCTION −g myplot myplot.i ...

declares that *myplot* calls graphical algorithms.

### 1.5.2 *Graphical Parameters*

In general, graphical functions should allow the user to supply settings for graphical parameters (line type, color, plotting character, etc.) as optional arguments. The recommended strategy is that the parameters should be changed before plotting is done, and restored (i.e., reset to the values which existed before the graphical function was called) after all plotting is complete. This does not apply to parameters which should stay in force to allow additional plotting on the current figure (e.g., axis parameters or user co-ordinate range). See the S function *par* [Becker and Chambers (1984), page 378] for a list of graphical parameters.

To pick up any graphical parameters in the arguments to a

function, include the special argument PAR; for example,

```
FUNCTION myplot(
    x/REAL/
    y/REAL/
    PAR
    )
```

PAR does not match a single argument in the usual way. Instead, it invokes an algorithm which scans the actual arguments for the name of any of the graphical parameters, extracts the corresponding values from the arguments and sets the graphical parameters (see also section 3.3.1). At the same time, the *previous* values of all these parameters are saved internally in the interface routine.

Upon exit from the graphics function, one of two actions will be taken. If the special structure PAR is returned, the graphical parameters will retain their new settings, and the previous values will be passed back as a special component of the result. This is used exclusively with the CHAIN statement. If *myplot* CHAINs to some other graphics function, for example *axes*, which also accepts graphical parameters, then the second function will leave the new parameter settings in effect during plotting and then restore the original values when it, in turn, exits. (This all happens automatically, so long as each of the graphics functions in the chain includes PAR as an argument and all but the last one in the chain execute "RETURN(PAR)".) The code is then of the form:

```
FUNCTION myplot(
    x/REAL/
    y/REAL/
    PAR
    &
    )

    ... #plotting stuff

    CHAIN(axes, PAR, FILTER)
END
```

Notice that we allowed extra arguments to *myplot* through "&", to be used by *axes* (such as *main* or *xlab*).

In the case that PAR is not returned, the original graphical parameters are automatically restored on exit from *myplot*. In any case, the parameters will be restored if an error or interrupt occurs during execution of the plotting function.

### 1.5.3 *Plotting Data Structure*

For many applications, a special *plotting data structure* is useful. This is a hierarchical structure with two components, named "x" and "y", corresponding to the x- and y-coordinates for a set of plotting positions. Generally, graphics functions that expect to get a set of plotting positions try to match any of the following possibilities:

- a plotting structure with components named "x" and "y";
- arguments *x* and *y* giving the two sets of co-ordinates;
- one argument to be interpreted as a time-series (either a true one, or a vector with "start" time 1).

To look for all of these possibilities automatically, use the statement PLOTARGS in place of the corresponding argument specifications. In *myplot* above, to allow for all these possibilities rather than just two vectors, the interface routine is written:

```
FUNCTION myplot( & )
STRUCTURE(x,y)
PLOTARGS
ARG( PAR, &)
... #and so on, as before
```

After PLOTARGS executes, vector structures *x* and *y* will be available for the x- and y-coordinate values respectively. The interface routine can go on to set up the plot (section 1.5.4) or can produce graphical output immediately, if the plotting is to be added to the existing plot.

By default, missing values are not allowed in PLOTARGS. If the application is prepared to deal with missing values, the interface routine should say:

```
PLOTARGS( NAOK )
```

### 1.5.4 *High-Level Graphics Functions: SETUP and LOGPLOT*

For graphics functions which produce a new plot, there are two interface language statements that automatically carry out typical setup requirements. The statement

```
SETUP
```

coming after structures *x* and *y* have been defined (typically by PLO-TARGS), will choose user co-ordinates and axis parameters to fit these values. In addition, SETUP will automatically look for arguments named *xlim=...* or *ylim=...*, and take these as overriding the axis limits implied by the data to be plotted.

The interface language statement

LOGPLOT

looks for a user-specified argument *log=*. It sets two static logical scalars, *logx* and *logy*, such that *logx* will be TRUE if the user specified a log transformation of the x-axis. For example if the user's call included an argument of the form *log="x"* or *log="xy"*, then *logx* will be TRUE. Similarly, *logy* will be TRUE if a log transformation of the y-axis was specified. LOGPLOT should be used before SETUP. When SETUP is executed, if either the x or y axes is on a log scale, the corresponding x or y variable is transformed by the *log10* function and a logarithmic axis is set up.

Thus, the paradigm for a high-level graphics function is:

```
FUNCTION myplot( & )
STRUCTURE(x,y)
PLOTARGS
LOGPLOT
SETUP
ARG( PAR, & )
 ... etc.
```

# 2
# Writing and Using Algorithms

Chapter 1 described the implementation of new S functions under the assumption that the subprograms (algorithms) were already available. In practice, however, one usually needs to write (or at least adapt) algorithms for the underlying computational problem at the same time. This chapter presents techniques for writing algorithms as well as a brief description of the algorithms supplied with S.

## 2.1 The Algorithm Language: Basics

While algorithms could be written in any form such that the resulting subprogram could be called from FORTRAN, the S software environment provides special features and tools to make the writing easier. Collectively, we refer to this as the S environment, and to the language used for writing algorithms (based on RATFOR and m4) as the *algorithm language*.

### 2.1.1 *Languages*

The language in which most of the algorithms supporting S are written is an enrichment of the RATFOR preprocessor for structured FORTRAN. The enrichment consists of some general features (described in this section) and a number of specialized facilities (section 2.2).

Files containing programs or subprograms developed in the algorithm language are identified by the suffix of the file names. Files ending in ".r" are compiled using RATFOR, those ending in ".e" are compiled with EFL, and those ending in ".C" or ".c" are compiled using C. The general and special facilities described below are available either with RATFOR or EFL. However, the facilities available with C are more limited (on the other hand, C provides some compensating facilities within the language). Section 2.1.6 describes the available C features.

### 2.1.2 *MAKE: Generating S Functions and Stand-Alone Programs*

Generally, it is not necessary to compile and load subprograms individually in the S environment. Instead, one associates files containing subprograms in the algorithm language with an S function, via the FUNCTION utility discussed in Chapter 1. The utility MAKE then generates an up-to-date version of the S function or program. An S function is declared by

$ S FUNCTION name file1 file2 ...

where *name* is the name of the S function and *file1*, etc. are files of source code in the interface language or the algorithm language. For example if S function *myreg* has an interface routine on file *myreg.i* and RATFOR-based supporting algorithms on files *newreg.r* and *comp2.r*, then the S function is declared (once) by

$ S FUNCTION myreg myreg.i newreg.r comp2.r

(See section 1.1.)

In this book, we emphasize developing algorithms in the S interactive environment. This greatly simplifies such time-consuming aspects of the process as generation of test data and analysis of results. However, it is also possible to generate stand-alone programs, for use within the UNIX operating system but outside of S, using the facilities of the S algorithm language. To declare a new stand-alone program, use the PROGRAM utility:

$ S PROGRAM name file1 file2 ...

where *name* is now the name you want for the executable program ("a.out" file). For example, suppose we want a stand-alone program test called *testreg* which tests *newreg.r* and *comp2.r*, and that the main program for the test is on *testreg.r*:

$ S PROGRAM testreg testreg.r newreg.r comp2.r

Once the FUNCTION or PROGRAM declaration has been completed for the S function or the stand-alone program, the actual program can be generated by

$ S MAKE name

with *name* either the S function or the executable program name. The MAKE command is a specialization and extension of the UNIX system command *make*. It carries out the compiling and loading steps as needed to produce the executable program. Note that the files on which *name* depends are not supplied to MAKE; this information has been stored away in a file named *Smakefile* by FUNCTION or PROGRAM. Whenever one of the files is updated, repeating the MAKE command will cause the executable version to be updated. For further details, see the comments in section 1.1.6 on the use of S MAKE.

If there is a change in the set of files on which *name* depends, the information used by MAKE must be updated by invoking FUNCTION or PROGRAM with the revised list of files. Note that FUNCTION or PROGRAM should not be redone when the *content* of files change, only when the list of files changes.

**2.1.3** *Error Handling*

Fatal errors, warnings, and messages to the user are written into algorithms by statements similar to those in the interface language:

FATAL(message)
WARNING(message)
MESSAGE(message)

What happens after the FATAL message is printed, however, is different  If the algorithm was called from an S function (i.e., directly or indirectly from an interface routine), the message from the algorithm will be followed by a message

Error in *name*

where *name* is the name of the S function. In the case of a stand-alone program, the FATAL message is followed by an exit from the program, with no further message. Error messages that need to print data values may use the ABORT statement, which prints all its arguments on the standard error file (see section 2.2.1) and then acts like FATAL.

**2.1.4** *Symbolic Constants; Declarations*

The algorithm language provides a number of constants in symbolic form. These increase readability. Also, since most of the constants are potentially machine-dependent, programming with them in symbolic form increases portability. The more frequently used constants, and their meaning, are:

| PRECISION | The smallest pos. real such that $1.+\text{PRECISION} > 1$ |
|---|---|
| BIG | The largest real |
| SMALL | The smallest positive real |
| BIGEXP | The largest exponent (base 10) |
| LARGEINT | The largest decimal integer |
| INT, REAL, CHAR | Modes |
| NCPW | The number of characters per (FORTRAN) word |
| NBPC | The number of bits per character |
| NDIGITS | The number of decimal digits to represent reals |
| EOS | The string terminator character |
| ESCAPE | The character to escape special characters |
| PI | 3.14159... |
| DEG2RD | PI/180. |

In addition to these constants, there are two special declarations:

> CHARACTER(name,width,dim1,dim2,...)

declares *name* to be a variable or an array to hold character data. (Because different FORTRAN implementations may have different formats for this, the CHARACTER statement aids portability.) The *width* gives the maximum number of characters in an element of *name*. The arguments *dim1*, etc. are optional and if given, are the dimensions of the array of character mode elements.

> POINTER p1,p2,...

declares variables or arrays *p1*, etc. to be pointers for use with dynamically allocated data (section 2.2.4). POINTER variables are actually integers as far as FORTRAN is concerned, but we recommend using the declaration to document the use of the variables as pointers.

There are also a number of less frequently used symbolic constants that appear in Appendix A.

**2.1.5** *Debugging*

When an algorithm is called from an S function, the DEBUG statement may be used inside the algorithm, as in the interface language (section 1.1.7). The interactive debugging facility will allow the user to query the contents of any of the S data structures in the interface routine. In order to use this feature, the interface routine must also contain some form of DEBUG statement. If no actual debugging is desired in the interface routine, the statement

DEBUG(ON)

can be used in the interface routine.

For stand-alone programs or for data which does not correspond to S structures, the printing facilities of section 2.2.1 may be used. There is no S facility to look at a symbol table for non-S structures, but interactive debugging facilities, such as *adb* or *sdb*, may be helpful.

**2.1.6** *C Language Facilities*

Unlike most UNIX application systems, S is primarily based on FORTRAN rather than C. However, C language subprograms may be used as support for S functions, provided the writer handles the interface between the C routine and the FORTRAN-based routines calling it. Some aspects of this interface are provided as S algorithm language utilities. The most important is the naming convention used to refer in C to FORTRAN subroutines and common blocks. Normally (but not always), *f77* modifies the names of globals to avoid conflict with similarly named C globals. The modification used depends on the version of the UNIX system. To ensure correct interpretation, C code that wants to refer, say, to the FORTRAN subroutine *mysub* or the common block *myblok* should refer to them as

F77_SUB(mysub)
F77_COM(myblok)

S will define the macros F77_SUB and F77_COM suitably for the local flavor of *f77*. Other local dependencies may exist in the FORTRAN/C interface (e.g., whether character variables in a subroutine call generate an extra argument giving the declared length). However, taking care of the naming convention as above generally makes writing C subprograms to support S functions reasonably straightforward.

The algorithm language facilities available to C-based routines are principally the symbolic constants of section 2.1.4 and the error handling routines of 2.1.3. The statements FATAL, WARNING, and MESSAGE are similar to their use in RATFOR or EFL, but slightly more general. The first argument is an unquoted format string, as for the UNIX routine *printf*. It is possible to include additional arguments, which will be interpreted as data items to be printed using the format string. For example:

> if( maxstep <= 0)
>> FATAL(Maxstep is %ld. It must be positive, maxstep)

(Avoid using commas in the string.) The example also illustrates that FATAL, WARNING, and MESSAGE are complete C language statements (no terminating semi-colon is needed).

The special algorithm-language facilities described in section 2.2 are not available to C-based routines. Note that C algorithms on ".C" files will be compiled with the S algorithm language facilities, but those on ".c" files will be compiled as pure C, with no specialization to S. The latter is the right choice for pure UNIX C programs, for which the writer does *not* want any of the S algorithm language features. A good programming practice in most cases is to have an outer layer of the C code that interfaces to FORTRAN (and used the S algorithm language features), and an inner layer that is pure C, independent of S.

## 2.2 Special Facilities

The algorithm language has a number of specialized facilities which are not made available automatically to all routines. When one or more of these facilities is needed, the programmer can make it available by putting the statement

> INCLUDE(name)

among the declarations of the subprogram. The *name* specifies the particular facilities wanted, as described in the subsections below; e.g., *print* for the printing facilities. The INCLUDE statement causes a set of specialized algorithm language statements to be made available and may also insert the declarations for some global variables (i.e., common blocks) at this point in the subprogram.

More than one facility can be included in a single program, by multiple arguments to INCLUDE, separated by commas, or by multiple INCLUDE statements.

### 2.2.1 *Printing and Encoding*

The algorithm language provides a facility for formatted printing and encoding of data. This replaces and extends FORTRAN-style formatted output. To have access to the printing facilities

INCLUDE(print)

should appear among the declarations of the program.

### 2.2.1.1 *Basic Message Printing*

The simplest and most common use of the facilities is to print some message, including data values, either as an error message or as part of the user's standard printed output. Printing to *standard output* is produced by:

PRINT($arg_1$, $arg_2$, ... )

where $arg_1$, etc. are items to be printed. Printing to *standard error* is produced by

EPRINT($arg_1$, $arg_2$, ... )

EPRINT is suitable for errors and other messages to the user; PRINT is for the printed display of results from the analysis. Formatted output can be combined with the generation of a FATAL error by the ABORT statement:

ABORT($arg_1$, $arg_2$, ... )

acts like EPRINT, followed by an error exit equivalent to that produced by FATAL. It is also possible to use ABORT with no arguments to exit after printing messages with EPRINT.

The arguments $arg_1$, etc. are either quoted strings or special format items (data values and optional formatting parameters) in the algorithm language. Data values are encoded according to their mode, either in free format or by a user-specified format. Stick to the simpler free-format items listed here until you need to produce particularly elegant print-out:

| "This is a message" | String |
|---|---|
| R(x) | Real value |
| I(jj) | Integer value |
| L(x>0) | Logical value |
| C(msg) | Characters with EOS |
| C(buff,length) | Chars with count |
| Q(msg) | Characters, printed in quotes |
| Q(buff,length) | Quoted chars with count |

Strings may contain any printing character except comma, "#" or unbalanced parentheses (see 2.2.1.4). The argument to R, I or L can be any expression of the appropriate mode. The printing facility examines the data value and chooses a suitable format. A leading space will be inserted to separate successive items, except character literals enclosed in quotes. For example:

> PRINT(I(nrow)," by",I(ncol)," matrix")

It is possible to direct printed output to a file other than standard output or standard error, by using

> FPRINT(ifile, arg1, arg2, ... )

instead of PRINT or EPRINT. Before FPRINT is used, the desired output file must have been opened. The integer *ifile* is the file descriptor returned by the opening routine *sopen* or *sattac* (see section 2.2.3).

Most S functions return a data structure and do no printing. However, there may be two kinds of output from algorithms to be used with S functions: error messages and displays. In general it is not a good idea to write new functions which produce printed displays unless there is really something new about the printing itself. Otherwise it is much cleaner to work with (possibly CHAIN to) one of the existing S functions for printing. Informative error messages, on the other hand, can often benefit from printing out the data values involved in the error. These can use EPRINT or ABORT just as above:

> ABORT("Iteration failed after",I(niter)," steps; still",
>         R(delta)," apart.")

If you really need to write a new display function, the output should be directed to the file OUTFC, rather than STDOUT, since the S user will then be able to redirect the output by using the *sink* function. The OUTFC file is opened automatically in all S functions, but an algorithm needs to include the definition of OUTFC by the statement

INCLUDE(io)

among the declarations. Also, any function using OUTFC must be declared using "FUNCTION −r" (section 1.1.6).

### 2.2.1.2 *Encoding*

The same facilities which produce printed output can also be used to encode strings. The encoded text can then be used for labeling of graphics, as part of an S structure or for any other purpose. The statement

ENCODE(arg1, arg2, ... )

encodes data items in the same way as the PRINT, etc. statements, but leaves the resulting text in a character variable, BUFFER. The integer variable BUFPOS gives the current length of the text in BUFFER and BUFLEN gives the maximum size of BUFFER. Multiple encode statements will concatenate the successive strings in BUFFER. The contents of the buffer may be printed and BUFPOS reset to 0 at any time by use of PRINT or EPRINT without arguments. For example

```
ENCODE("Statistics:")
do i=1,5
        ENCODE(R(stats(i)))
PRINT
```

This illustrates one convenience of the ENCODE statement: building up a line of output iteratively. Normally, one will need to check that the buffer has not become too full.

```
PRINT("High:")
do i=1,nhigh {
        if(BUFPOS>LINEWIDTH−FIELDWIDTH) PRINT
        ENCODE(R(x(i)))
        }
if(BUFPOS>0)PRINT
```

The last line prints any fields on the final line of output. Symbolic constants FIELDWIDTH and LINEWIDTH specify the maximum length of an encoded numeric field, and the longest line of printed output.

The other main application of encoding is to make character string data items. The usual tool for this purpose is the supplied algorithm:

istrng(text,length)

which returns a POINTER value to a stack copy of *length* characters of *text*. For example,

ENCODE("Group",I(ngroup))
ip=istrng(BUFFER,BUFPOS)
CLEAR

generates the encode text in BUFFER and then assigns to *ip* a pointer to a copy of the same text, on the dynamic stack. If encoding is not followed by a PRINT or similar statement, it is essential that the buffer be cleared by the CLEAR statement (otherwise the encoded string will appear at the front of the next PRINT message).

Note that the text in BUFFER does not have a string terminator at the end, unless it is provided explicitly; e.g.,

ENCODE("Group",I(ngroup),"EOS")

After "Group" and the integer field are encoded, the special character EOS is inserted (EOS is a symbolic constant, not three separate characters).

### 2.2.1.3 *Detailed Format Control*

When detailed control of the printed output is desired, spacing and specific format information can be given. The following arguments to PRINT, ENCODE, etc. control spacing:

| | |
|---|---|
| SP(n) | Space forward n characters |
| T(m) | Tab to position m |
| TM(iw) | Tab to a multiple of iw |

The TM argument is used to print output in columns *iw* wide. The T operation tabs to column *m*, or inserts 1 blank, if the current BUFPOS is greater than *m*. As an example of spacing, consider the "high" values in the example above as a display indented 7 characters.

```
ENCODE("High:")
do i=1,nhigh {
        if(BUFPOS>LINEWIDTH−FIELDWIDTH) {
                PRINT
                ENCODE(T(7))
                }
        ENCODE(R(x(i)))
        }
```

```
if(BUFPOS>7)PRINT; else CLEAR
```

Each line after the first starts in position 8; the test for printing the last line is modified correspondingly.

Data formats may be given explicitly in the R, I and L fields. The I and L fields take an optional width argument:

```
ENCODE( I(ij,5) )
```

encodes an integer field (right-justified) 5 characters wide. REAL fields take three parameters: width, number of positions after the decimal point and a third argument of E_FORMAT or F_FORMAT to select the format type. Typically, these format values are specified in order to give the same format for a set of real numbers. In this case, the format parameters should be chosen by an algorithm, *rdtfmt*. The programmer need not worry about them personally. For example,

```
call rdtfmt(x,nhigh,i1,i2,i3)   #compute format for x
ENCODE("High")
do i=1,nhigh {
    ...
    ENCODE( R(x(i),i1,i2,i3) )   #use format
}
```

modifies the previous example to use the same format for all values. The algorithm *rdtfmt* takes a vector of real values and its length, and returns the three format parameters.

One further specialization of the formats is to give a length of 0. This is equivalent to encoding in free format, except that the leading space which separated items is suppressed. Suppression of internal spaces is a good idea, for example, if the resulting string becomes the name of a dataset or component:

```
ENCODE("Gr",I(ngroup,0))
ip=istrng(BUFFER,BUFPOS)
CLEAR
```

makes character strings of the form "Gr5" or "Gr1066" for corresponding values of *ngroup*.

### 2.2.1.4 *Problems with Encoding*

Because the printing and encoding facilities are implemented by the *m4* macro processor, characters that have a special meaning to *m4* must be encoded by special mechanisms; they can not be included in strings. The four characters most commonly needed are comma, sharp and parentheses (left and right). For these there are four special format items:

    COMMA
    SHARP
    L_PAREN
    R_PAREN

It is *not* possible to print any of these characters in the middle of a literal string (enclosed in quotes).

    PRINT("The three values are",I(i1),COMMA,
        I(i2),COMMA," and",I(i3))


### 2.2.2 *Reading and Decoding*

Facilities are provided for input of data items and the breaking up of text into fields corresponding to data values. These correspond in a symmetric way to the printing and encoding facilities of 2.2.1. To give a program access to the reading facility:

    INCLUDE(read)

This activates definitions of the reading and decoding operations and also provides access to a buffer for reading, INBUF, along with its current and maximum length, INPOS and INLEN. The reading and printing buffers are separate, so that reading and printing can go on concurrently within a single subprogram.

### 2.2.2.1 *Basic Reading of Data Items*

The fundamental viewpoint of the reading facility is that input comes as a stream of characters. Fields within the stream are separated by "white space" (blanks, tabs or new lines). To read a number of fields from the standard input stream (typically the user's terminal):

    READ( arg1, arg2, ... )

A line will be read from the input file. The fields described by *arg1*, etc. will be decoded from the line in the mode requested. If the end of the line is reached, another line will be read, and so forth.

The fields are described similarly to printed fields:

| R(x) | Real |
|------|------|
| I(ii) | Integer |
| L(flag) | Logical |
| C(buffer,maxlen) | Characters |
| S(buffer,maxlen) | Characters with EOS |

The first argument to each field must be a variable or array element, since the decoded value will be stored in it. The two character fields differ only in whether a string terminator character is inserted at the end of the field. Notice that the argument is a character buffer (not a pointer) and that a maximum length should be supplied to protect against overflowing the buffer. Other than this, no lengths or format control should appear in the decoding statements. There are alternative ways to read data when it is in fixed-format records. See, for example, Becker and Chambers (1984), section 5.4.4 and the documentation for *extract* in S for a utility which extracts data from a file, given column formatting information.

A typical use of the READ facility might be to read user input in response to a prompt, for example:

```
MESSAGE(Enter values for n and x)
READ( I(n), R(x) )
```

The user can supply input on one line or several; the READ statement will not complete until both fields have been decoded. Errors in interpreting fields produce a FATAL error message. See 2.2.2.2 for line-oriented input and end-of-file detection.

It is not necessary for the entire input to be read in one statement. The statement

```
DECODE(arg1, arg2, ... )
```

will decode additional fields, starting just after the last decoded field. If there are no more fields on the current line, the next line of input will be read. For example, to read first the number of items and then all the items, allowing as many of the items as desired to be on any line:

```
READ( I(n) )
for(i=1; i<=n; i=i+1) DECODE( R(x(i)) )
```

READ does not need any arguments: when called without arguments, it fills the input buffer, ready for any DECODE statements. Note that using a second READ statement after the first will throw away any unused fields in the current line. For example, suppose we want to read character fields from the input, but skip the rest of a line after a field which starts with "#".

```
repeat {
        DECODE( C(field,maxlen) )
        if(jgetch(field,1) == "#" ){
               READ
               next
               }
     ...
```

The *jgetch* function retrieves the first character from the field (see section 2.3.1).

### 2.2.2.2 *Line Input; End-of-File*

Fields can also be decoded from input which is read a line at a time.  This form is useful if a specific number of fields are to be taken from each line.  To get a line of text into the input buffer,

> GETLINE(ifile)

where *ifile*, if supplied, is the integer file descriptor for a previously attached file (section 2.2.3).  If *ifile* is omitted, input is from the standard input (normally the terminal).

The following example reads pairs of real numbers from each line, until end-of-file is reached.

```
for( i=1; i<=imax; i=i+1) {
        GETLINE(ifile); AUTO_CHECK=FALSE
        DECODE( R(x(i)) )
        if(EOF)break
        DECODE( R(y(i)) )
        if(EOF)break
        }
```

If GETLINE encounters an end-of-file condition, it sets the internal EOF flag, which should be checked by the calling program, as above. Setting AUTO_CHECK to FALSE says that the program will do explicit EOF checking; otherwise the first item DECODED will produce an automatic error in case of end-of-file.  The DECODE following GETLINE will not read further lines.  When the last field from the current line has been decoded, further DECODE requests will set EOF and immediately return.

It is possible to decode strings which were not originally read from a file; for example, strings which were arguments to an S function. The statement

GETSTRING( text )

moves the string *text* into the decoding buffer. Subsequent DECODE
statements then operate as after the GETLINE statement.

### 2.2.3 *File Access; Standard Files*

The S environment provides the routines for opening and closing files, to use with the printing and reading operations of the previous sections. Files can be opened in an algorithm or interface routine
by

ifile=sopen(name,perms)

where *name* is the name of the file (a character string terminated by
the end-of-string character) and *perms* is READ, WRITE, or
READWRITE for read-only, write-only, or read-write permission. The
function *sopen* (remember to declare it integer) returns a positive
integer *file descriptor* if the open was successful and a negative integer
if the open failed (e.g., the file could not be opened or permission was
refused). When the file input/output is complete, the file is closed by

call sclose(ifile)

In S interface routines or algorithms called by them, a more automatic
procedure for attaching files can be used. This puts the files on a list
of open files, which can be automatically detached if appropriate. To
use these facilities,

INCLUDE(attach)

must appear in the interface routine or algorithm. To attach a file

ifile=sattac(name,perms,type)

where *name* and *perms* are as before and *type* specifies when the file is
to be automatically detached: values of AUTO, ONERROR, and PERM
specify that the file disappears automatically at the end of the expression, on encountering an error and never, respectively. Also,
remember to declare *sattac* integer. In any case, the file can also be
detached explicitly by

call sdetac(ifile)

Generally, it is bad strategy to attach and detach files in an S
function unless there are special requirements. It is normally much
easier to convert files to data structures in S before the individual S

function sees them (using *read* or *?extract*, for example).

In any case, try to ensure that attached files are faithfully detached, either through the AUTO feature in an S function or by explicitly using *sdetac*.

### 2.2.4 *Dynamic Storage*

The S environment allows dynamic allocation and release of blocks of storage. This is the preferred way to provide working space to numerical and other algorithms. To use the dynamic allocation facility

INCLUDE(stack)

should appear in the subprogram. Storage is allocated by the function *jstkgt*. This returns a *pointer* to dynamically allocated storage sufficient to hold a specified number of data items of a specified mode; e.g.,

ip=jstkgt(n,REAL)

returns a pointer to space for $n$ items of mode REAL. All variables used as pointers (and functions returning pointers) to the stack should be declared POINTER:

POINTER ip,jstkgt

When storage has been allocated on the stack, elements of the allocated storage can be referenced in the appropriate mode:

```
rs(ip)                          # REAL
is(ip)                          # INTEGER
ls(ip)                          # LOGICAL
```

For example, to allocate $2*n$ integers and initialize to $0,...,2n-1$

```
iptr=jstkgt(2*n,INT)
for(i=0; i<2*n; i=i+1) is(iptr+i)=i
```

When the dynamic storage is no longer needed, the $k$ most recently allocated blocks of storage may be released by

call jstkrl(k)

Note that storage is allocated in a *last-in-first-out* form. It is not possible to release a dynamically allocated block unless you also release all blocks allocated later than the one in question.

**2.2.5** *Data Structures for S*

This section is very advanced, and is intended for those writing S support algorithms which require knowledge of S data structures. It assumes familiarity with these structures, as discussed in section 3 of Appendix B.

Algorithms (not written in the interface language) which need to refer to the structure of S data should have

INCLUDE(struct, stack)

in the routine.

Whether in an algorithm or in an interface routine, suppose that *ptr* is a pointer to an S data structure. In an algorithm, *ptr* will be passed in through the argument list or will be the result of previous computations with S structures. All S data structures will have the attributes

NAME(ptr)
MODE(ptr)
LENGTH(ptr)

for the (pointer to) the name of the structure, the (integer) mode of the structure and the (integer) length (number of data elements). Every structure also has the pointer attribute

VALUE(ptr)

which points to the values (on the stack of dynamically allocated data) of the data structure.

If *ptr* points to a vector, then the mode will have one of the values REAL, INT, LGL or CHAR. In the first three cases, the data values for the vector are found on the corresponding stack. If

n=LENGTH(ptr)
v=VALUE(ptr)

then the data items are

rs(v), ..., rs(v+n−1)        # reals
is(v), ..., is(v+n−1)        # integers
ls(v), ..., ls(v+n−1)        # logicals

If the mode is CHAR, then *v* points to a vector of *n* pointers, each of which points in turn to the actual text of the corresponding string. For example, to print each of the character strings using the printing facility of section 2.2.1:

```
for(i=0; i<n; i=i+1)
     PRINT( Q(TEXT(is(v+i))) )
```

Note that the *integer* stack is used to refer to an array of pointers. Similarly, to allocate an array of pointers, use INT as the mode argument to *jstkgt*.

If the original structure was not a vector, but a hierarchical structure, then MODE(ptr) is STR. In this case, VALUE points to a *directory* whose entries are the components of the structure. Expressions in the algorithm language allow reference to the entries: suppose that *dirptr=VALUE(ptr)* where *ptr* points to a structure. Attributes defined for directories include:

| | |
|---|---|
| FIRSTENT(dirptr) | Pointer to the first component |
| LASTENT(dirptr) | Pointer to the last component |
| ENTRY(dirptr,j) | Pointer to the j-th component |

A typical operation on a hierarchical structure is to loop over all the components, performing some computations on the component data structures. One way to do this is

```
dirptr=VALUE(ptr)
for( i=1; i<=LENGTH(dirptr); i=i+1) {
        compnt=ENTRY(dirptr,i)
        ... # now do the work on compnt
        }
```

Note that *compnt* is a pointer to the substructure, so that all the attributes (NAME(compnt), MODE(compnt), etc.) are defined.

The loop over the components shown above allows computations over one level of substructures. It is possible to loop over all the components of a structure at any level. This process is usually called a depth-first search or tree traversal. This process is less often needed than simply looping over one level of components, and requires some knowledge of tree structures and of the *m4* macro processor.

S provides for a depth-first scan of any hierarchical structure. To use the scan

```
INCLUDE(tree)
```

should appear in the program. The scan moves over all components of the structure. As each component is encountered, its mode is tested. If the mode is STR (the component is a structure itself) the scan then moves DOWN to examine all subcomponents. Otherwise the scan moves ALONG to the next component at the current level.

Finally, when the scan has gone over all the components at the current level, it moves UP to the next component at the level above.

The algorithm language statement

>     TREEWALK(ptr)

where *ptr* is the pointer to an S data structure, carries out the scan. The individual routine defines up to three *actions*; namely, DOWN, ALONG, and UP. Each of these is specified by defining a macro of the given name, containing the statements to be executed. The definitions must appear before the TREEWALK statement.

The attribute CURRENT is the pointer to the structure component currently being examined. Other attributes defined are LEVEL, the current depth within the structure, the pointer PARENT, the structure above CURRENT, the pointer NEXT, the (directory of) the structure to which the tree-walk will go down, and MAXLEV, the maximum depth of structure allowed.

A very simple example prints the level and the name of all components:

```
subroutine trpr(ptr)
POINTER ptr

INCLUDE(tree,print)
define(`DOWN',
   `EPRINT( I(LEVEL), C(TEXT(NAME(CURRENT))) )')
define(`ALONG',
   `EPRINT( I(LEVEL), C(TEXT(NAME(CURRENT))) )')

TREEWALK(ptr)
EPRINT("End of structure")
return
end
```

## 2.3 Available Algorithms

This section presents brief discussions of algorithms which may be of general interest. Omitted from this discussion are most of the algorithms which are close analogues of individual S functions. Look at the source code for the S function to see what algorithms are used, if you want an algorithm for a similar purpose. The on-line S help command:

> **help("System")**

will tell you how to find relevant S source code. (This information also appears in Appendix C.)

### 2.3.1 *Data Handling; Character Data*

Arrays of all modes may be filled with constant values or copied into other arrays.

```
call rfill(data,x,length)                    # REAL data
call ifill(idata,ix,length)                  # INTEGER data
call lfill(ldata,lx,length)                  # LOGICAL data
```

These all fill arrays with the single data item given. To copy data,

```
call rcopy(xfrom,xto,length)
call icopy(ifrom,ito,length)
call lcopy(lfrom,lto,length)
```

handle REAL, INTEGER and LOGICAL vectors.

The utilities to fill and copy character data are:

```
call cfill(char,buffer,length)
call concat(to,ito,from,ifrom,length)
```

Copying character data is handled somewhat more generally, since one may wish to specify the starting character position in one or both strings. The routine *concat* copies *length* characters from the *ifrom*-th character of *from* to *to*, starting at the *ito*-th character. The character positions are numbered from 1.

Two other basic operations for moving character data are provided:

```
ich=jgetch(string,i)
```

gets the *i*-th character from *string*. Note that both *ich* and *jgetch* should be declared

```
CHARACTER(ich,1)
CHARACTER(jgetch,1)
```

The opposite process of putting a character into a string is

```
call jputch(string,i,ich)
```

There are a number of algorithms and macros to convert among various forms of character data and pointers to strings.

TSTRING(anything)

puts an EOS on the end of some literal text and encloses it in quotes. To convert text to a pointer to a string on the stack

ptr = istrng(text, length)

If *length* is given as −1, the text will be scanned for a terminating EOS character. If character data is initially packed into a buffer in the form of *n* fields, each *length* characters long, the data may be converted into a vector of *n* pointers to character strings on the stack:

ptr = ch2vec( buffer, n, length)

This returns a pointer to *n* pointers on the integer stack, the *i*th pointer pointing to a character string made from the *i*th field in *buffer*. Trailing blanks are removed from each field, and an EOS character appended.

To test the equality of two strings, use the logical function *streq* (remember to declare it):

if(streq(text1,text2)){ ... *matched* ... }

See also the partial match techniques used in S (section 1.2.2).

### 2.3.2 *Sort and Order*

There are algorithms for sorting data or producing the ordering permutation. These are analogues of the S functions *sort* and *order*, with the crucial difference that in the FORTRAN-based algorithm language, there must be different versions of the routines for different modes. For sorting,

call sort(x,length)          # REAL data
call sorti(ix,length)        # INTEGER data

sort vectors of REAL or INTEGER data. There is then a general sort

call sortf(ix,length,fun)

In this routine, *fun* is the name of an external integer function. This function will be called by the sort routine to compare two data items from the vector to be sorted. The function *fun* takes two integer or POINTER arguments and returns −1 if the first argument should precede the second argument, 0 if they are equal, and 1 if the first argument should follow the second argument. The usefulness of doing the compares in a function is that any ordering can be followed, not just the obvious numerical one.

The common example is that of comparing and thus sorting

character strings. The algorithm *chcomp(ip1,ip2)* compares two strings, given two pointers *ip1* and *ip2*. This leads to the following use of *sortf* to sort a vector of (pointers to) character strings.

```
subroutine sortc(ptr,length)
POINTER ptr(length); integer length

external chcomp

call sortf(ptr,length,chcomp)
return
end
```

Note that data of mode POINTER is actually of FORTRAN mode integer.

Parallel to the sorting routines are routines which return, in addition to the sorted data, a vector of integers giving the ordering permutation (see the S function *order*).

```
call orderr(x,length,iorder)
call orderi(ix,length,iorder)
call orderf(ix,length,iorder,fun)
```

In each case, *iorder* is an integer vector of *length* items, returned filled with the permutation which orders the vector *x* or *ix*.

### 2.3.3 *Range of Data*

For plotting purposes, one needs to compute the range of data values.

```
call rangev(x,length,xmin,xmax)
```

returns in *xmin, xmax* the minimum and maximum value of the vector *x*. Given several sets of data, one can obtain the cumulative minimum and maximum of all the sets by the function *rangec*. For example, suppose *x1*, *x2* and *x3* are three vectors.

```
call rangev(x1,n1,xmin,xmax)
call rangec(x2,n2,xmin,xmax)
call rangec(x3,n3,xmin,xmax)
```

The value returned from *rangec* for *xmin* will be the minimum of the data values in the argument and the input value for *xmin* and similarly for *xmax*.

If the data may have NAs included, the range of all the actual values may be obtained from call narang(x,n,xmin,xmax)

Note that other algorithms discussed in section 2.3 *do not* allow NAs.

### 2.3.4 *Probabilities; Quantiles; Pseudorandom Numbers*

As explained in Becker and Chambers (1984), section 3.5, probability distributions have been assigned code names, and functions to produce probabilities, quantiles, and random values have names formed from "p", "q" and "r" prepended to the code. The underlying algorithms have similar names, but produce only a single value as result.

#### 2.3.4.1 *Available Algorithms*

Not all of the distributions are represented by explicit algorithms; random values from the remaining distributions are computed by the inverse probability law; i.e., *qDIST(runif( ... ))* is equivalent to *rDIST( ... ))*. Random number generators are provided explicitly for the following distributions:

| | |
|---|---|
| chis | Chi-square |
| expos | Exponential |
| gamm | Gamma |
| normk | Normal |
| pois | Poisson |
| stab | Stable family |

The calling sequence for each of these includes all parameters *except* for location and scale parameters, followed by an dummy seed argument.

xchis=rchis(idf,iseed)

returns a Chi-squared random value with *idf* degrees of freedom.

#### 2.3.4.2 *New Pseudorandom Generators for S*

*This is an advanced topic.* For the most part, new facilities for generating pseudorandom numbers in S can and should be done by generating uniform or other random numbers and transforming these according to the generating scheme desired. Typically, this is a good application for an S macro.

If a new function is to be written which generates random numbers, either for internal use or to return them, it must mesh with the technique used by S to make all random numbers flow from one consistent sequence of values (and, in particular, to avoid having the S function start its generator at the same place every time). S uses a special dataset, named "Random.seed", to keep the seed values for the generators. This dataset is read in at the beginning of each function generating random values, and written out at the end.

Any S function that generates random numbers should be declared with the "−r" option of the FUNCTION utility (see the documentation). Before computing any pseudorandom values, the interface routine should include the statement:

GETSEED

After computing all desired pseudorandom values, the interface routine should include:

PUTSEED

All new random generators should use the generator *uni* as an underlying source of uniform random numbers.

**2.3.5** *Matrices and Arrays*

Numerous algorithms are included for numerical computations on matrices. There are also two routines for in-place transposition of matrices.

call transr(x,nrow,ncol)
call transi(ix,nrow,ncol)

where *x* and *ix* are real and integer matrices with *nrow* rows and *ncol* columns.

The following numerical routines are of general use. Most of them require some knowledge of numerical linear algebra, to apply the algorithms in new areas. We give only the names here; see the source code for argument lists.

| | |
|---|---|
| backsl, backsu | Back-solve triangular system |
| chol | Choleski decomposition |
| condu | Estimate condition number |
| eigen1 | Eigenvalues |
| dot, dotv, dotwv | Inner products |
| gs | Q-R decomposition |
| gsls, gslsi, gslsw gslsiw | Least-squares |
| matp, matpt | Matrix product |
| svd | Singular-value decomposition. |

There are other routines associated with individual S functions, either with corresponding names or mentioned in the S function documentation.

# 3
# Graphical Algorithms

This chapter discusses the creation of graphical algorithms, for use within S functions. Graphical algorithms are based upon a set of device-independent graphical subroutines designed for use in data analysis. The algorithms built from these subroutines are device-independent, and can communicate with many different graphical devices through device driver programs.

The underlying graphical subroutines range from general, high-level routines to specific, low-level routines. For example, some subroutines produce entire plots, complete with axes and titling. At the other extreme are subroutines that draw a line or plot a text string. Intermediate level routines are available for tasks such as setting up an axis or plotting a set of titles. This structure provides building blocks for constructing new graphical algorithms.

A large set of graphical parameters allows control over special characteristics of the resulting plots (color, character size, etc.). At the same time, the graphical subroutines provide reasonable default values for parameters that the algorithm writer does not wish to specify. In this case, the algorithm *user* (either the user of an S function incorporating the algorithm or the author of a subroutine calling the algorithm) retains the ability to change these parameters. Strategy parameters control more general aspects of the plotting, such as the overall shape of the plot (e.g., square, maximum sized for the device), the layout of multiple figures on a single page, and methods used to generate pretty axis labels. The current value of any parameter can be specified or queried by graphical algorithms. Plotting is done in specialized coordinate systems: the major portion of the plot is carried out in the *user* coordinate system, titling and axis labeling in a *margin* coordinate system.

Section 3.1 describes the basic concepts of the graphical subroutines; 3.2 describes the construction of a graphical algorithm, using an example; 3.3 extends the example to both the S and stand-alone environments; 3.4 treats more advanced details of the graphical subroutines and parameters. A number of subroutines are described in section 3.5 and section 3.6 is concerned with device drivers.

## 3.1 Basic Concepts of Graphical Algorithms

There are three basic concepts which are necessary for you to understand the S graphics facility:

- the way in which graphical output is structured into *figures* containing labeling information which surrounds a *plot* of data-analytic information.
- the relevant *coordinate systems* in which graphical operations are expressed.
- the use of *graphical parameters* to control the graphical output.

Familiarity with these simple concepts provides the basis for understanding how to work with and write graphical algorithms.

### 3.1.1 *Figures and Plots*

A typical plot consists of a central portion containing the primary graphical information (a scatter plot, curves, a histogram), surrounded on four sides by auxiliary information: titles, axis scaling, legends, etc. We will refer to the central region as the *plot*, and the border surrounding it as the *margins*. Together, the plot and margins make up a complete *figure*. This is illustrated in Figure 1.

The figure is a complete piece of graphical output, and occupies all or part of a page of output or the surface of a graphical device. A device may be organized to display one or more figures. The *outer margins* border the edge of the page, and provide an area for overall titling when more than one figure is present. The region inside the outer margins may be divided up among one or more figures (see Figure 2).

### 3.1.2 *User and Margin Coordinate Systems*

The natural coordinate system for the plot region and its information is in the units of the user's data. For example, a map may be plotted in degrees of latitude and longitude, a plot of height vs. weight in inches and pounds, etc. These coordinates are known as

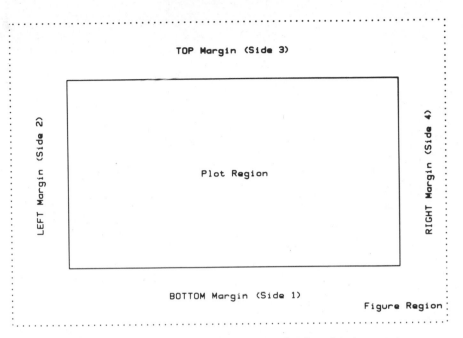

**Figure 3.1.** The figure region is composed of a plotting region surrounded by margins.

*user* coordinates.

The titling information, however, is often more conveniently expressed by its relationship to the plotting region. Thus, a title may be centered two lines above the plot. The *margin* coordinates are given by the side (bottom, left, top, or right), the distance from the corresponding edge of the plot, and a position along the side. Since the information plotted in the margins is most often text, it is convenient to describe the distance out from the edge of the plot in lines of text. The position along the side of the plot can be described in several ways, either as a fraction of the way along the side, or by the corresponding user coordinate. Margin coordinates are also used in the outer margins.

There are other coordinate systems that are occasionally useful. It is possible to relate the user coordinate system to the physical dimensions of the device, e.g., to ensure desired spacing or size for graphical output. To explicitly position a figure on the page, it is useful to specify the positioning in fractions of the way up or across the page. It is also possible (though unusual) to give the plot position explicitly as a fraction of the figure.

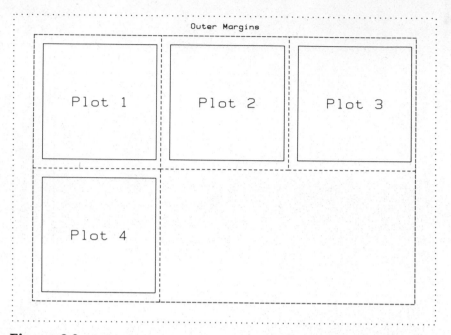

**Figure 3.2.** Multiple figures on a page: each contains a plot region and margins, and all figures are surrounded by outer margins.

### 3.1.3 *Graphical Parameters*

Perhaps the most important concept in the construction of graphical algorithms is that of *graphical parameters*. The precise effect of a graphical operation is not defined until a number of decisions have been made by the graphical system. For example, in plotting a scatter of points, the graphical system must decide upon the character (or symbol) to plot at each point, what size the character should be, its rotation, color, etc.

The graphical subroutines use a set of graphical parameters to determine, at a given time, exactly how to carry out specific operations. Values of the parameters may be queried or specified by a graphical algorithm, or may be left at their current values. Reasonable default values exist initially for all parameters. Parameters *not* modified by an algorithm may still be specified by a higher-level routine or by the user of an S function.

In order for the concept of allowing outside control of graphical parameters to work, graphical algorithms must follow a strict discipline whenever changing the parameter values: the parameter value is queried and saved, it is changed, the operation using the changed value is carried out, and the saved value of the parameter is then restored. Restoration of the old parameter value prevents the parameter setting from having a permanent effect. The only parameters changed without restoration are those relating to coordinate systems and axes, specified by high-level algorithms. These parameters are not restored, since that would prevent adding further information to the current plot.

## 3.2 Creating Graphical Algorithms

For the most part, graphical algorithms are constructed by the techniques described in section 2. They are generally written in the *algorithm language* (based upon the RATFOR structured FORTRAN preprocessor), and utilize facilities provided by the S algorithm environment. This section describes techniques that are unique to the writing of graphical algorithms.

Suppose that we are interested in writing an algorithm to produce a special form of scatter plot, in which each point is marked by a cross of variable size. For example, the cross may be used to indicate the magnitude of errors in measuring the x and y variables.

We would like to have

call cross(x,y,n,dx,dy)

produce a plot with $n$ crosses $(x_i \pm dx_i, y_i \pm dy_i)$. The remainder of this section will construct this algorithm while introducing the basic operations for graphical algorithms.

### 3.2.1 *Initialization of Graphical Algorithms*

The statement

INCLUDE(graphics)

should appear in the declaration section of the algorithm (along with type statements, common declarations, etc.). This enables the graphical parameter facility, which is later accessed by the two operations QUERY and SPECIFY, described in section 3.2.5.

Algorithms like *cross* that produce a new plot, must also initialize the graphical system at execution time by means of

call beginz

This routine has two effects: it interacts with the device driver to receive the current values of the graphical parameters, and it causes an advance to a clear figure on the device. If no device driver is currently in effect, *beginz* will cause a fatal error. If no plotting has yet been done on the device, *beginz* will establish all parameters at their default values.

Thus far, our routine looks like:

```
#cross  scatter plot with crosses at x+/−dx, y+/−dy
subroutine cross(x,y,n,dx,dy)
integer n
real x(n),y(n),dx(n),dy(n)

INCLUDE(graphics)

call beginz
```

### 3.2.2 *Setting-up Coordinate Systems and Axes*

Once the graphical subroutines have been initialized, it is necessary to set up a coordinate system appropriate to the plot. For this, we need first compute the minimum and maximum values to appear on each axis. In many cases the algorithms *rangec, rangev,* or *narang* (section 2.3.3) can be used to find the range of data to be plotted. However, for this plot we need to find the extremes of the crosses, and it would be unduly conservative to assume that the largest cross goes with the most extreme data points. Thus, for the x-axis of *cross*:

```
xmax=x(1)+dx(1); xmin=x(1)−dx(1)
for(i=2; i<=n; i=i+1){
        xmin=amin1(xmin,x(i)−dx(i))
        xmax=amax1(xmax,x(i)+dx(i))
        }
```

Now, given *xmin* and *xmax*, the statement

```
AXIS(BOTTOM,xmin,xmax)
```

will set up a coordinate system that includes at least the values from *xmin* to *xmax*. Also, graphical parameters are stored to define the "pretty" values at which the axis should be labeled. The axis is not drawn yet; this is normally done after the plot is finished (section 3.2.4).

A similar computation is then done for the y-axis

```
... compute ymin, ymax ...
AXIS(LEFT,ymin,ymax)
```

(See section 3.4.4 for information about setting up logarithmic or time axes, or for details of how labels are chosen.)

### 3.2.3 *Drawing the Picture*

With the coordinate system set up appropriately, we can now begin the actual plotting of the crosses. The calls

```
for(i=1; i<=n; i=i+1){
    call segmtz(x(i)+dx(i), y(i), x(i)−dx(i), y(i), 1)
    call segmtz(x(i), y(i)+dy(i), x(i), y(i)−dy(i), 1)
    }
```

draw the sets of horizontal and vertical line segments used in our plot.

As in this example, it is often straightforward to draw the desired picture once coordinate axes have been set up. The basic sub-routines for plotting data are:

```
call pointz(x,y,n)
call linesz(x,y,n)
call segmtz(x1,y1,x2,y2,n)
call arowsz(x1,y1,x2,y2,n)
```

which plot a set of $n$ points, draw connected line segments through a set of points, or draw a set of $n$ non-connected line segments (or arrows) from the points $(x1(i),y1(i))$ to the points $(x2(i),y2(i))$. All of the x- and y-coordinates are expressed in the previously established user coordinate system. In addition, the routine

```
call crclsz(x,y,r,n)
```

draws a set of circles.

It is frequently useful to place text on the plot. The routine

```
call textz(x,y,string)
```

plots a single character string *string* at the user coordinate position $(x,y)$. The string is centered by default. *String* should be terminated by an end-of-string (EOS) character, as described in sections 1.2.2 and 2.3.1.

Even the most complicated plots are normally produced by suitable combinations of these basic point, line, arrow, circle, and text plotting routines. Section 3.5 describes a number of routines that produce higher level graphical objects through the use of the basic routines described in this section.

**3.2.4** *Titles and Axis Labels*

With the body of the plot completed, it is only necessary to provide suitable labels. First, the coordinate axes (set up in section 3.2.2) are labeled by

call saxisz(side,ticks,labels)

The argument *side* takes on the values BOTTOM, LEFT, TOP, or RIGHT. Arguments *ticks* and *labels* are logical values (TRUE or FALSE) specifying whether the tick marks and/or the numeric labels for the axis should be drawn.

For the cross plot,

call saxisz(BOTTOM,TRUE,TRUE)
call saxisz(TOP,TRUE,FALSE)  #only ticks on top
call saxisz(LEFT,TRUE,TRUE)
call saxisz(RIGHT,TRUE,FALSE)  #only ticks on right

Titles can be placed in the margins surrounding the plot by

call mtextz(side,line,string)

Once again, *side* is BOTTOM, LEFT, TOP, or RIGHT. Real-valued argument *line* tells the number of lines of text to leave between the edge of the plot and the title. Together, *side* and *line* give a margin coordinate. *String* is a character string (terminated by an end-of-string character) that is centered and plotted parallel to the specified side of the plot.

Thus, to place a descriptive label on our plot:

call mtextz(BOTTOM,4.,
    TSTRING(Cross Represents Error for Each Observation))
call boxz

TSTRING creates a terminated string, and *boxz* produces a box surrounding the plot.

**3.2.5** *Specifying and Querying Graphical Parameters*

This section gives some examples of the use of graphical parameters [Becker and Chambers (1984), pages 92-95] within graphical algorithms. Section 3.4.5 will give a summary of all the available parameters. In the cross-drawing example, suppose that we would like character-string identifiers at each point on our cross plot. The identifiers should extend from the center of the cross up and to the right at a 45 degree angle. The calling sequence for the routine changes to

```
subroutine cross(x,y,n,dx,dy,id)
integer n
real x(n),y(n),dx(n),dy(n)
POINTER id(n)
```

(See section 1.2.2 for a discussion of vectors of pointers to character strings.)

The basic subroutine *textz* can be used to plot the strings. However, we must alter several graphical parameters to make the text appear as intended. First, string rotation must be set to 45 degrees, rather than the default 0 degrees (horizontal). Second, the string should be left-justified, not centered. This can be accomplished by

```
SPECIFY( srt(45), adj(0) )
```

to specify the graphical parameters *srt* (String RoTation) and *adj* (string ADJustment). Rotations are measured in degrees counter-clockwise from horizontal. Adjustment varies from 0. (left-justified), through 0.5 (centered), to 1. (right-justified).

If we left-justify the strings at the center of the cross, we will find that the first character overplots the cross. We would like to move out from the center by some multiple of the character size. To do this, query the current character size in user-coordinates:

```
QUERY( cxy(cx,cy) )
```

Now the identifiers can be placed on the plot by

```
SPECIFY( srt(45), adj(0) )
QUERY( cxy(cx,cy) )
for(i=1; i<=n; i=i+1)
      call textz(x(i)+cx,y(i)+cy,TEXT(id(i)))
```

One remaining problem is that once the string rotation and adjustment parameters have been changed from their default values, any successive text plotting will no longer have the default values of parameters *srt* and *adj*. As mentioned above, the paradigm is to remember the old values of parameters that are to be changed, to set the new values, and finally to restore the parameters to their old values.

```
QUERY( srt(oldsrt), adj(oldadj) )
SPECIFY( srt(45), adj(0) )
... text plotting here ...
SPECIFY( srt(oldsrt), adj(oldadj) )   # restore parameters
```

Parameters are generally named by 3-letter mnemonic strings. As in the example, their values are specified or queried by

SPECIFY( parm(values), parm(values), ... )
QUERY( parm(variables), parm(variables), ... )

Here *parm* is the mnemonic parameter name, *values* is a list of values that the parameters should take on, and *variables* is a list of variables that are to be set with the current values of the parameters.

There are a number of parameters available for dealing with text besides *srt* and *adj*. Character rotation within a string is controlled by parameter *crt*. By default, this parameter is equal to the most recently specified value of *srt*; thus to have characters rotated at a different angle than the strings it is important to specify *crt* after *srt*.

The size of characters is generally also adjustable, and is controlled through the parameters *cex* (Character EXpansion relative to standard size), *csi* (Character Size in Inches), and *csr* (Character Size Relative to the size of the figure region). Parameters *cex* and *csr* are normally recommended, since *csi* depends more on the size of the output device. For example, to make characters which are double the standard size

SPECIFY( cex(2) )

QUERY can be used to make characters double their *current* size:

QUERY( cex(cursiz) )
SPECIFY( cex(2*cursiz) )

With all of the character-oriented parameters, the precise results depend on the graphics device used at execution time. Certain devices have only discrete sets of allowable character sizes and rotations. In general, the device drivers will attempt to make the display hardware obey the graphical parameters, but they will not take extraordinary measures to do so. (Characters are not drawn line-by-line in order to make their size and rotation continuously variable).

There are a number of other parameters that affect the basic drawing subroutines of section 3.2.3. For line drawing, *lty* takes on integer values (dependent on the particular device) to specify various dotted and dashed Line TYpes. Parameter *lwd* controls Line WiDth.

Both lines and text are affected by the *col* parameter, which, on devices with the capability, causes color changes.

The full set of graphical parameters and their default values is presented in section 3.4.

**3.2.6** *Wrapping Up*

The last graphical call in a high-level algorithm should be

call finisz

This signals the end of the algorithm, and causes the device driver to note any final changes to parameter values.

With this call, we now have our algorithm

```
#cross  scatter plot with crosses at x+/-dx, y+/-dy
subroutine cross(x,y,n,dx,dy,id)
integer n; real x(n),y(n),dx(n),dy(n)
POINTER id(n)

INCLUDE(graphics,stack)

call beginz

xmin=x(1)-dx(1); xmax=x(1)+dx(1)
for(i=2; i<=n; i=i+1){
     xmin=amin1(xmin,x(i)-dx(i))
     xmax=amax1(xmax,x(i)+dx(i))
     }
AXIS(BOTTOM,xmin,xmax)

ymin=y(1)-dy(1); ymax=y(1)+dy(1)
for(i=2; i<=n; i=i+1){
     ymin=amin1(ymin,y(i)-dy(i))
     ymax=amax1(ymax,y(i)+dy(i))
     }
AXIS(LEFT,ymin,ymax)

for(i=1; i<=n; i=i+1){
     call segmtz(x(i)+dx(i), y(i), x(i)-dx(i), y(i), 1)
     call segmtz(x(i), y(i)+dy(i), x(i), y(i)-dy(i), 1)
     }

QUERY( srt(oldsrt), adj(oldadj), cxy(cx,cy) )
SPECIFY( srt(45), adj(0) )
for(i=1; i<=n; i=i+1)
     call textz(x(i)+cx,y(i)+cy,TEXT(id(i)))
SPECIFY( srt(oldsrt), adj(oldadj) )

call saxisz(BOTTOM,TRUE,TRUE)
call saxisz(TOP,TRUE,FALSE)
call saxisz(LEFT,TRUE,TRUE)
call saxisz(RIGHT,TRUE,FALSE)

call mtextz(BOTTOM,4.,
     TSTRING(Cross Represents Error for Each Observation))
call boxz
```

```
call finisz
return
end
```

With some artificial data, it produces Figure 3.

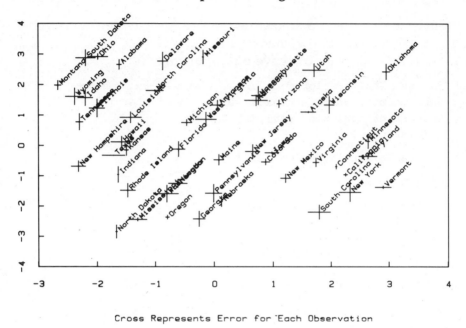

Cross Represents Error for Each Observation

**Figure 3.3.** Plot produced by the *cross* algorithm.

## 3.3 The Structure of a Graphical Algorithm

In section 3.2, we constructed a graphical algorithm *cross* to produce a new form of high-level graphical figure. In this section, we will put it to use.

As it was presented in section 3.2, *cross* is appropriate as a high-level algorithm; that is, as part of an S function that generates a complete plot. In this section we first discuss how such a function might be written (section 3.3.1) and then consider how the algorithm might be modified to augment a plot (section 3.3.2).

### 3.3.1 *High-Level Graphical Algorithms; S Functions*

The first step is to declare the new function via the FUNC-TION utility described in section 1.1.2. Remember that all graphics functions must be declared with the option "−g". To create an S function *cross*, type

> S FUNCTION −g cross cross.i cross.r

In order to turn an algorithm into an S function, it is necessary to write an interface routine for it. As described in section 1.5, the S interface language provides a number of facilities for the construction of graphical functions. Specifically, the PLOTARGS and SETUP statements provide a convenient mechanism for recognizing a variety of arguments to graphical functions and for setting up axes. PLOTARGS allows (x,y) data to be recognized either as two separate arguments or as a single graphical data structure (section 3.4.4).

```
FUNCTION cross(&)
STRUCTURE(x,y)
PLOTARGS      # gets (x,y) data
ARG(
        dx    /REAL/
        dy    /REAL/
        id    /CHAR/
        PAR
        &
        )
```

The additional argument PAR allows user-specified parameter values to be given as arguments to this function, and the final argument "&" allows other (unused) arguments.

By default, the SETUP statement automatically uses the range of arguments *x* and *y* to set up limits for the axes. (It also calls *beginz* to signal the start of a high-level graphical routine.) We would like, instead, to use a computation that allows room for the crosses. To accomplish this, we can use the optional arguments

> SETUP(xmin,xmax,ymin,ymax)

At this point, it becomes obvious that a subroutine to compute our axis limits is a good idea.

> call crange(x,dx,LENGTH(x),xmin,xmax)
> call crange(y,dy,LENGTH(y),ymin,ymax)
> SETUP(xmin,xmax,ymin,ymax)

where *crange* is the routine

```
#crange   compute range of x+/−dx
subroutine crange(x,dx,n,xmin,xmax)
integer n; real x(n),dx(n),xmin,xmax
```

```
xmin=x(1)−dx(1); xmax=x(1)+dx(1)
for(i=2; i<=n; i=i+1){
     xmin=amin1(xmin,x(i)−dx(i))
     xmax=amax1(xmax,x(i)+dx(i))
     }
return
end
```

Now that the coordinate system is set up, we want to plot the crosses and identifiers. A subroutine constructed out of the inner layer of *cross* is useful for this purpose.

```
#cross2    produce cross plot with current coord system
subroutine cross2(x,y,n,dx,dy,id)
integer n; real x(n),y(n),dx(n),dy(n)
POINTER id(n)

INCLUDE(graphics,stack)

for(i=1; i<=n; i=i+1){
     call segmtz(x(i)+dx(i), y(i), x(i)−dx(i), y(i), 1)
     call segmtz(x(i), y(i)+dy(i), x(i), y(i)−dy(i), 1)
     }

QUERY( srt(oldsrt), adj(oldadj), cxy(cx,cy) )
SPECIFY( srt(45), adj(0) )

for(i=1; i<=n; i=i+1)
     call textz(x(i)+cx,y(i)+cy,TEXT(id(i)))

SPECIFY( srt(oldsrt), adj(oldadj) )

return
end
```

The interface routine can then call *cross2* to carry out the actual plotting.

The CHAIN statement takes more of the burden from the graphical algorithm by allowing another function to label the coordinate axes. The statement

CHAIN(axes,PAR,FILTER)

causes the *axes* function to draw the coordinate axes and surrounding box, and also to inherit the user-specified parameter values, and any arguments that this function did not use.

The final version of the *cross* interface routine is

```
FUNCTION cross(&)
STRUCTURE(x,y)
PLOTARGS
ARG(
```

```
            dx    /REAL/
            dy    /REAL/
            id    /CHAR/
            PAR
            &
            )

if(LENGTH(x)!=LENGTH(y)|LENGTH(x)!=LENGTH(dx)|
        LENGTH(x)!=LENGTH(dy)|LENGTH(x)!=LENGTH(id))
            FATAL(Lengths do not match)

call crange(x,dx,LENGTH(x),xmin,xmax)
call crange(y,dy,LENGTH(y),ymin,ymax)
SETUP(xmin,xmax,ymin,ymax)

call cross2(x,y,LENGTH(x),dx,dy,id)

CHAIN(axes,PAR,FILTER)
END
```

Notice that a check on the lengths of the arguments has also been included.

The most useful graphical algorithm for the S environment, then, is one that assumes that a coordinate system and axes have been set up by the interface routine. The algorithm draws the desired plot, and the interface routine CHAINs to function *axes* to label the plot.

In a stand-alone environment (outside of S), a graphical algorithm still needs to set up and draw axes. Therefore, when an algorithm is to be used both in and out of S, it is often constructed in two layers. The outer layer, used in stand-alone mode, sets up the axes, calls the inner routine, and labels the axes. The inner layer actually constructs the plot. When used in S, only the inner algorithm is called, with the facilities of the interface language replacing the outer routine. The stand-alone version of *cross* becomes

```
#cross   scatter plot with crosses at x+/-dx, y+/-dy
subroutine cross(x,y,n,dx,dy,id)
integer n; real x(n),y(n),dx(n),dy(n)
POINTER id(n)

INCLUDE(graphics)

call beginz

call crange(x,dx,n,xmin,xmax)
AXIS(BOTTOM,xmin,xmax)

call crange(y,dy,n,ymin,ymax)
AXIS(LEFT,ymin,ymax)
```

```
call cross2(x,y,n,dx,dy,id)

call saxisz(BOTTOM,TRUE,TRUE)
call saxisz(TOP,TRUE,FALSE)
call saxisz(LEFT,TRUE,TRUE)
call saxisz(RIGHT,TRUE,FALSE)

call mtextz(BOTTOM,4.,
    TSTRING(Cross Represents Error for Each Observation))
call boxz

call finisz
return
end
```

**3.3.2** *Algorithms that Augment a Plot*

In S, all graphical operations are carried out through a *device driver* that knows how to control a specific device. This operation is illustrated by the Figure 4.

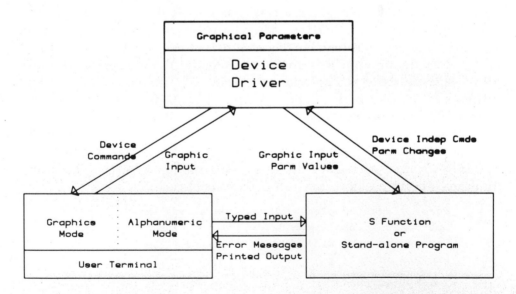

**Figure 3.4.** The operation of graphical device drivers, showing interactions between the device driver, the user's graphic terminal, and the device-independent graphic S function.

Once it is activated, a device driver communicates with all graphical functions and also maintains the current state of the graphical parameters. Because a device driver may interact sequentially with a number of different graphical algorithms, it is possible for one algorithm to set graphical parameters and for another algorithm to make use of these new parameter settings in later plotting. This is especially useful when one algorithm sets up a user coordinate system, and another algorithm augments that plot using the same coordinate system.

An S function that uses the algorithm *cross2* to augment an existing plot with crosses is

```
#add crosses to existing plot
FUNCTION addcross(&)
STRUCTURE(x,y)
PLOTARGS
ARG(
        dx      /REAL/
        dy      /REAL/
        id      /CHAR/
        PAR
        )
if(LENGTH(x)!=LENGTH(y) ... # as before

call cross2(x,y,LENGTH(x),dx,dy,id)

END
```

Notice how neatly the inner layer of our high-level algorithm can be used for this purpose.

## 3.4 Advanced Graphical Algorithms

Sections 3.2 and 3.3 should provide enough information for the construction of simple graphical algorithms. It is usually best to develop simple algorithms initially, and to refine them later.

This section describes in much more detail the activities carried out by the graphical subroutines. It is recommended for advanced readers.

### 3.4.1 *Control of Figures, Plots, Margins*

A number of graphical parameters are devoted to the positioning of figures on the output device and of plots within the figure. This section describes these parameters, and related graphical subroutines.

As Figure 3.2 illustrated, the device surface is divided into outer margins and figures, and each figure is divided into margins and a plot. Understanding this structure is important in writing algorithms that alter the size of any of these regions.

In general, a new S function *should not* change parameters dealing with these regions, but instead should produce a suitable plot in the currently defined plot region, and label it in the current margins. The S user can control the shape and arrangement of plots through the function *par* [Becker and Chambers (1984), pages 378-384].

Because it is desirable to allow S to determine margin size and figure placement, the parameters and subroutines described in this section should ordinarily be used only by algorithms that produce a number of figures or those with special needs. With this warning in mind, we now describe the parameters.

In the simple case, there is one figure on the device surface. By default, the figure occupies the entire surface. Within the figure are the margins. They are controlled by the parameters

    mar(bottom,left,top,right)
    mai(bottom,left,top,right)

The second, *mai*, refers to margin size in inches. The first, *mar*, measures the size of the margins by the number of lines of text that can be accommodated. The parameter *mex(size)* is a measuring unit for the margins which describes the size of these lines of text, relative to the standard character height (actually, the inter-line spacing) on the device. Parameter *mex* does not affect character size, it only changes the way in which the margin is allocated and addressed. Thus

    SPECIFY( mex(1.5), mar(3,4,5,1) )

allocates enough bottom margin to accommodate 3 lines of text of size 1.5, etc. The statement

    SPECIFY( mar(4.5, 6, 7.5, 1.5) )

allocates the same amount of margin space but does not change the units of the margin coordinate system which are measured in lines of *mex*-sized text. The margin coordinate system will be described further in section 3.4.2.

Once the space for margins is reserved in the figure region, the remainder of the space may be used for the plot. The shape of the plot depends on the parameter

    pty(type)

where *type* can be "s" for square or "m" for maximum size within the confines of the margins (the default). For special applications, the plot can also be specified in inches, by its aspect ratio, or as a fraction of the figure region by the routines *inpltz*, *arpltz*, and *drpltz*.

The situation becomes more complicated when more than one figure appears on the device surface. In this case, outer margins are normally allocated at the extremes of the device surface. The size of the outer margins is controlled by parameters

oma(bottom,left,top,right)
omi(bottom,left,top,right)

which, like their margin counterparts *mar* and *mai* specify outer margins in lines of *mex*-sized text or inches. By default, the outer margins are all of size zero. Outer margins can also be used when there is only one figure per page: for example, to provide consistent labeling for several pages, some with one figure per page and others with more than one.

Figures are allocated from the region not used for the outer margins. Most commonly, users want a regular array of figures on the device. The routines

call mfrowz(rows, cols)
call mfcolz(rows, cols)

set up a *rows* by *cols* array of figures, with *mfrowz* specifying that they are to be filled row-by-row, and *mfcolz* column-by-column. When in this mode, the parameter

mfg(i,j,m,n)

describes that the current figure is the $i$th row and $j$th column of an $m$ by $n$ array. This parameter can also be specified to begin plotting at a specific figure of the array. After the use of *mfrowz* or *mfcolz*, calls to *beginz* advance to the next clear figure in the array of figures.

Precise control over figure placement is available by subroutine *drfigz*, which directly specifies figure placement as a fraction of the device surface. In all of the above cases, parameter *fty* holds the type of the current figure, and the value "r" is multiple by rows, "c" is multiple by columns, "m" is maximum size, "d" is directly specified.

At any given moment, the graphical subroutines attempt to keep the graphical parameters consistent with one another. For example, if a change is made to the outer margins, the following sequence of operations takes place:

1 — The size of the outer margins is subtracted from the size of the device.

2 — The current figure is positioned within the remaining area according to the figure type and the *mfg* parameters.

3 — The size of the margins is computed from parameters *mar* and

*mex* and the known size of a standard character, and the margins are thus allocated within the figure.

4 — The plot is recomputed based upon plot type *pty*.

Changes to figure parameters start the computation at step 2. Changes to margin parameters cause computations to begin at step 3.

At any intermediate stage, the set of parameter values may be inconsistent. For example, suppose that we want to set up a 3 by 3 array of figures and try

call mfrowz(3,3)

In step 2 above, the current figure will have an area of 1/9 of the device surface. Then, margins will be allocated from this area. If the margin size has not been changed from its default value, there may be no room in this small figure for all of the margins, thus causing an error. This problem can sometimes be solved by cutting down on margin size before changing to multiple-figure mode, and increasing margin size only after returning to one figure per page. The rule of thumb is to make changes which would increase the size of the plot (or decrease the margin size) first.

Because the plotting region may be changed when related parameters are changed, figure or plot defining parameters should not be reset if they are changed by an algorithm. Resetting the parameters would recompute the plot region, etc., and thus prevent further algorithms from augmenting the plot.

### 3.4.2 *Margins and Outer Margins*

As described in section 3.1.2, text in the margin region is addressed by means of a margin coordinate system. This system uses three values to position text: *side*, *line*, and one *user coordinate*. The *side* value is either BOTTOM, LEFT, TOP, or RIGHT. For text positioned parallel to the side of the plot, the *line* value measures the distance from the edge of the plot to the near edge of the text. This measurement is done in units of *mex*-sized characters; i.e., if parameter *mex* is 2, then the units of the coordinate system are based upon the height of a twice-normal character and *line* 2.5 is two-and-a-half of those units from the edge of the plot. The user coordinate measures distance along the specified side.

This system provides a flexible means of addressing the margin region. The routine

> call mtextz(side,line,text)

is most often used to plot text in the margin. It utilizes the parameter *adj* to control the user coordinate position along the side: if *adj* is .5, the text is centered at the middle of the side, 0. left justifies at the minimum user coordinate, and 1. right justifies at the maximum user coordinate. The text is always plotted parallel to *side*.

A somewhat more general routine is

> call gtextz(side,line,coord,text)

which plots *text* at the specified margin coordinate using the current values of string rotation and adjustment.

By default, the operation of *clipping* is carried out at the edge of the plotting region. Any points, text, or lines that extend into or beyond the margin are truncated at the point where they leave the plotting region. This prevents graphical operations from cluttering the margin, and can be used to keep outlying points off a plot. Clipping can be modified to allow plotting to eXPanD into the margin by means of parameter *xpd*. If *xpd* is TRUE, clipping will be done at the edge of the figure, rather than the edge of the plot. This is particularly useful for graphical functions that augment current plots, like the *addcross* function presented in 3.3.2. To allow added crosses to extend into the margin,

> logical oldxpd
>
> ...
>
> QUERY( xpd(oldxpd) )
> SPECIFY( xpd(TRUE) )
> call cross2( ... )
> SPECIFY( xpd(oldxpd) )

The outer margins can be addressed identically to the margins. Routine *otextz* plots in the outer margin in a way analogous to *mtextz*. More generally, the parameter Outer Margins On:

> omo(on)

controls operations in the outer margin. If *on* is TRUE, any operations that ordinarily would have occurred in the margin are instead done in the outer margins. In addition, parameter *omo* preserves the user coordinate system, so that data specified in user coordinates appears in the correct place. This means that axes can be plotted in the outer margins by *saxisz*. For example, when plotting a number of figures on the same page, all having the same x axis, a single set of axis labels could be placed at the bottom of the page:

SPECIFY( mar(0,5,1,5), oma(5,0,5,0) )
call mfrowz(4,1)
 ... produce 4 figures separated by only 1 line of TOP margin ...
 ... do not use saxisz to draw any x axes ...
SPECIFY( omo(TRUE) )
call saxisz(BOTTOM,TRUE,TRUE)   #label x axis on bottom plot

### 3.4.3 *Parameters of Physical Size*

Since the device driver knows the physical characteristics of the graphical device, it is possible to QUERY parameters

pin(width,height)
fin(width,height)
din(width,height)

which give the width and height of the plot and figure regions, and the device surface, respectively.

Another useful parameter is

uin(ux,uy)

which gives the number of inches corresponding to one user coordinate unit in both the x- and y-directions. This parameter is often used when physical relationships are to be preserved or when plotting is to be done in a specific distance. For example, to position a character string 1 inch to the right and 1 inch above the point $(x,y)$

QUERY( uin(ux,uy) )
text( x+1./ux, y+1./uy, TSTRING(My Label) )

It is also important occasionally to ensure that the x- and y-values of *uin* are equal. In drawing a picture of an object, it is appropriate to scale the size of the object to fit on the page, but not to change the x and y scales independent of one another. To set up a coordinate system addressed in inches,

QUERY( pin(px,py) )
SPECIFY( usr(0,px,0,py) )

To make sure that the x- and y-axes are commensurate, it is possible to make the plot type parameter square, and set up the coordinates identically on each axis:

call rangev(x,n,umin,umax)
call rangec(y,n,umin,umax)

SPECIFY( pty("s"), usr(umin,umax,umin,umax) )

### 3.4.4 *Setting-up Coordinate Systems and Axes*

Once a plot and figure have been set up, it is normally necessary to establish a user coordinate system. The default user coordinate
system places the point (0,0) at the lower left hand corner of the plot
region, and (1,1) at the upper right. These default coordinates are
sometimes reasonable to use in graphical algorithms, but most often,
the algorithm will be more natural if an application-specific coordinate system is used.

If the algorithm knows the exact range of the user coordinates
and there will be no need to generate axes, the user coordinates can be
specified directly:

SPECIFY( usr(xl,xu,yl,yu) )

Be sure not to try to draw axes with *saxisz*, etc., since the axis parameters will not be correctly defined.

Since most applications require the plotting of coordinate axes,
section 3.2.2 introduced the AXIS statement to set up a linear axis and
coordinate system easily. The remainder of this section will present
routines for detailed control of axes.

First, however, we should mention several graphical parameters that affect axes. For example,

tck(size)

determines the length of axis TiCK marks as a fraction of the plot size.
The default value is 0.02; interesting variations include negative
values (ticks point outside of plot) and the value 1.0 (tick marks
become grid lines).

Parameter

lab(nx,ny)

suggests the number of intervals that should appear on axes set up by
high-level routines. Parameter *las* tells how axis labels should be
oriented (0=parallel to the axis, 1= always horizontal) and *mgp* positions the axis line and labels in the margin.

*The remainder of this section is very advanced, and can be ignored by
most readers.* Subroutines which set up the coordinate system and
coordinate axes at the same time are:

```
call stdaxz(side,style,umin,umax,nint,flag)
call logaxz(side,style,umin,umax,nint,flag)
call timaxz(side,style,umin,umax,nint,flag)
```

to set up standard (linear) axes, logarithmic axes, and time axes, respectively. Note that the $x$ and $y$ axes are set up in separate calls, allowing arbitrary combinations of standard, logarithmic, and time axes.

Argument *side* is BOTTOM, TOP, LEFT, or RIGHT. (BOTTOM and TOP refer to $x$ axes, LEFT and RIGHT to $y$ axes). The arguments *umin* and *umax* are the minimum and maximum user coordinates desired on the axis, and there should be approximately *nint* intervals labeled along the axis.

Before describing the *style* and *flag* arguments to these routines, it will be valuable to discuss how labels are chosen for axes. In general, a plotted axis is labeled with "pretty" numbers; namely, numbers which are 1, 2, or 5 times a power of ten. The distance between adjacent labels is another "pretty" number.

The axis routines operate by first finding a "pretty" number which is near the value *(umax−umin)/nint*. This value becomes the distance between axis labels. Then, pretty numbers for the ends of the axis are chosen. The *style* parameter controls this choice. In most cases, the axis is labeled at values more extreme than *umin* and *umax*. For example, if these arguments were 1.357 and 7.68, the axis might range from 0 to 8 in steps of 2, or from 1 to 8 in steps of 1. This "standard" axis labeling is used if *style* is "s". It may also be desirable to make maximal use of the page for the plotted data. In this case, an "internal" axis can be used. Here, pretty values of the axis extremes are chosen *within* the values of *umin* and *umax*. In the example, the axis could be labeled from 2 to 7 in steps of 1.

The internal axis style brings up an important point: the axis limits and the corresponding coordinate system limits are related, but are not necessarily identical. For an internal axis, the coordinate system ranges from *umin* to *umax*, but the axis is labeled at pretty values within that range.

Another axis style is "extended". The axis labels are identical to those of a standard axis, but the user coordinates are extended to ensure that neither *umin* nor *umax* lie right at the user coordinate limit. This prevents data from being plotted directly on the box surrounding the plot. Similarly, "rational" axis style extends the coordinate limits, but by a fixed 7% beyond the end of the data. This ensures that the displayed data will occupy a fixed fraction of the plotting region.

If *style* is given as " ", then the previous style is used. This again gives the higher-level routine a measure of control over the lower-level operations. The default style is "e".

Logarithmic and time axes are parameterized in a way similar to linear axes. In all, each axis is described by:

1 — *min*imum and *max*imum user coordinates (parameter *usr*).
2 — the *type* of axis: standard (linear), logarithmic, or time (parameters *xaxt* and *yaxt*).
3 — the *style* of axis labels: standard, internal, rational, or extended (parameters *xaxs* and *yaxs*).
4 — a set of three parameter values telling where axis labels should be placed. For linear axes, these three values are the minimum axis label, the maximum axis label, and the number of intervals the axis contains. Logarithmic and time axes are also described by three values (parameters *xaxp* and *yaxp*).

We can finally describe the *flag* argument to these routines. If *flag* is 3, all of the axis parameters are computed as outlined above, their values are saved in the graphical parameter array, the user coordinate system is modified, and the axis is then plotted. If *flag* is 2, the plotting is suppressed. (It is often best to produce a plot and label the axes last, thus giving the user time to look at the display while the axes are drawn.) Standard, logarithmic, and time axes can be plotted later by routines *saxisz*, *laxisz*, or *taxisz*. Finally, if *flag* is 1, everything but setting the user coordinates is performed. This is used for special applications, for example, where only part of the user coordinates are to have labeled axes. Another routine that performs this task is *diruxz*.

### 3.4.5 *Summary of Graphical Parameters*

The following table gives a list of the mnemonic abbreviations for the graphical parameters, along with their default values. The parameters are grouped according to the types of graphical output they affect.

| What | Units or Values | par(Values) | Default |
|------|-----------------|-------------|---------|
| **Characters** | | | |
| Size | Relative to standard | cex(size) | 1 |
| Size | Inches | csi(inches) | − |
| Size | Fraction of Plot | csr(fraction) | − |
| Size | User coords | cxy(width,height) | −† |
| String Rotation | Degrees ccw from horiz | srt(angle) | 0 |
| Char Rotation | Degrees ccw from horiz | crt(angle) | 0 |
| String Justification | 0=left, 1=right, .5=center | adj(pos) | .5 |

**Lines**

| | | | |
|---|---|---|---|
| Line Type | device dependent | lty(type) | 0 (solid) |
| Line Width | device dependent | lwd(width) | 1 |

**Other Symbols**

| | | | |
|---|---|---|---|
| Plotting Char | character | pch(char) | "*" |
| Mark Height | inches | mkh(inches) | same as std char |
| Circle Smoothness | rasters | smo(error) | 1 |

**Lines, Characters**

| | | | |
|---|---|---|---|
| Color | device dependent | col(color) | 0 |

**Margins**

| | | | |
|---|---|---|---|
| Outer Margins | lines of text | oma(bot,lef,top,rig) | 0,0,0,0* |
| Outer Margin | inches | omi(bot,lef,top,rig) | 0,0,0,0* |
| Outer Margins | fraction of device | omd(xl,xu,yl,yu) | 0,1,0,1 |
| Outer Margins On | logical | omo(on) | FALSE |
| Margins | lines of text | mar(bot,lef,top,rig) | 7,7,7,7* |
| Margins | inches | mai(bot,left,top,rig) | — |
| Margin Units | fraction of standard | mex(size) | 1 |

**Plots and Figures**

| | | | |
|---|---|---|---|
| Figure Type | character | fty(char) | "m" (max size) |
| Multiple Figures | array m by n | mfg(i,j,m,n) | 1,1,1,1 |
| Clipping | logical | xpd(off) | FALSE |
| Plot Type | character | pty(char) | "m" (max size) |
| Box Type | character | bty(char) | "o" |
| New Plot | logical | new(empty) | TRUE |

**Axes**

| | | | |
|---|---|---|---|
| Axis Type | character | xaxt(char) | "s" (standard) |
| | | yaxt(char) | |
| Axis Style | character | xaxs(char) | "e" (extended) |
| | | yaxs(char) | |
| Axis Parameters | label info | xaxp(p1,p2,p3) | 0,1,5 |
| | | yaxp(p1,p2,p3) | |
| User Coordinates | | usr(xl,xu,yl,yu) | 0,1,0,1 |
| Label Style | 0=parallel, 1=horiz | las(style) | 0 |
| Tick Length | fraction of plot | tck(leng) | 0.02 |
| Label Intervals | number | lab(nx,ny) | 5,5 |
| Axis Position | lines | mgp(lab,num,line) | 3,1,0 |

**Size**

| | | | |
|---|---|---|---|
| Device Size | inches | din(width,height) | −† |
| Figure Size | inches | fin(width,height) | −† |
| Plot Size | inches | pin(width,height) | −† |
| Coordinate Size | inches per unit | uin(ux,uy) | −† |
| Raster Size | inches | rsz(rx,ry) | −† |

**Miscellaneous**

| | | | |
|---|---|---|---|
| Error Mode | −1=ignore, 0=print | err(mode) | 0 |
| Device Code | device dependent | dev(code) | −† |

Unless marked with †, all parameters can be both specified and queried. In general, the parameters listed in the table can be specified or queried with as many arguments as are of interest. Thus, for example, to specify only the x-coordinates

SPECIFY( usr(xl,xu) )

or to QUERY only the y-coordinates

QUERY( usr(,,yl,yu) )

A few parameters (indicated by an "*" in the table) can not be abbreviated in this way for SPECIFY. The effect of parameters listed as "device dependent" can be found by looking at the documentation for the individual S device driver functions in Becker and Chambers (1984).

### 3.4.6 Graphical Input

Most graphical devices provide some special form of *graphical input*. The way in which it is carried out differs from device to device: pen plotters normally have buttons or a joystick to allow the user to manually position the pen, scopes may have cursors or mice, etc. For devices with no special mechanism, the device driver will ask the user to type in the desired coordinates. Details can be found in the documentation for the S device driver functions. Whatever the mechanism, the device driver takes care of appropriate user interaction in response to the subroutine

call inputz(x,y,n,nmax)

and returns up to *nmax* different pen positions in vectors *x* and *y*. The value of *n* gives the number of points that were actually given by the user. The returned points are all in user coordinates.

This graphic input facility is used in the S functions *rdpen* and *identify*, and could be used in other algorithms to interact with a menu of commands, identify outliers, etc.

### 3.4.7 *Debugging*

Debugging is often an easier process for a graphical algorithm than for a purely computational one. The ability to watch the interactive graphics being produced often provides clues to the origin of problems. Of course, any of the debugging techniques of section 1.1.7 and 2.1.5 can be used with graphical algorithms.

The values of the graphical parameters are often useful for debugging. They can be obtained by the S function *pardump* which takes as arguments the character string names of graphical parameters. Thus

        pardump("usr","srt")

will give the current user coordinate and string rotation values. The function *pardump* with no arguments gives the values of all parameters.

A similar facility is available in subroutine form. Executing

        call dumpaz

Specific parameters can be printed by querying them and then using the algorithm language printing facilities (section 2.2.1).

## 3.5 Available Graphical Subroutines

A number of high-level graphical subroutines are available for use in graphical algorithms. We list some of them here; see the S source code for the detailed calling sequences. We have left them out of the discussion until now because most of these high-level routines are already available in the form of S functions. For example, high-level routines exist for contour-, scatter-, and box-plots.

| barz | Bar plot |
|---|---|
| bplotz | Barycentric plot of mixture x+y+z data |
| bxpz | Box plots |
| contrz | Contour plot |
| eepltz | Empirical q-q plot |
| eqpltz | Empirical-theoretical q-q plot |
| hhpltz | Histogram |
| nplotz | Scatter plot with repeated point counts |
| persp | Perspective plot |
| piez | Pie chart |
| plot1 | Very high-level controlled scatter plots |
| pltusa | Map of USA |
| rplotz | Robust scatter plot |
| splotz | Scatter plot |
| tspltz | Time-series plot |

More interesting, perhaps, are the routines that underly these high-level subroutines. These often produce the essence of the plot, without dealing with details of axis construction, labeling, etc. For example, the routines *bxz* and *bxnz* draw a single standard or notched box for use in a boxplot. Underlying routines like this may be useful in the construction of graphical algorithms.

Some of the potentially useful routines are:

| bhtchz | Shades bars of bar chart |
|--------|--------------------------|
| bmglgz | Legend for shaded bar chart |
| bnamez | Names for bars of bar chart |
| bxz | Box to represent distribution |
| bxnz | Notched box for distn with conf limits |
| cont2z | Contour lines |
| hatch | Shade in polygonal area |
| hdlinz | High-density vertical lines |
| hhbrkz | Break points for histogram |
| hhdonz | Number of classes for histogram |
| linesp | Lines with transformation set up by pictur |
| lintfz | Lines with transformed coordinate system |
| npntsz | Points with counts of repeats |
| pictur | Set up translation, scale change, rotation |
| pnttfz | Points with transformed coordinate system |
| ptitle | Titles |
| rpntsz | Robust plot of scatter of points |
| shadez | Shade in rectangle |
| titlez | Titles |
| tsvecz | Vector of x values for time-series plot |
| usabdy | Coordinates of boundary of USA |

## 3.6 Device Drivers

A number of S device drivers are available (often in the stand-alone environment as well). They include the Advanced Electronics Design 512 color scope, Hewlett-Packard 2623, 2647, and 2648 scopes, Hewlett-Packard 2627 color scope, Hewlett-Packard hpgl (7470, 7475, 7220) pen plotters, Hewlett-Packard 7221 pen plotter, Ramtek 6211 color scope, Tektronix 4006, 4010, 4012, and 4014 scopes, Tektronix 4112 (and other 4110 series) scopes, Tektronix 4662 pen plotters, and printer-like devices (printer). There are also device drivers for UNIX system graphic devices, and the *pic* typesetter graphics program.

Even with this variety of device drivers available, new graphic devices are continually being manufactured. This section describes how users can construct their own device drivers for new graphical devices.

**3.6.1** *Organization of Device Driver Routines*

A device driver is constructed from a number of individual subroutines, each of which carries out a well-defined task. Specifically, there are routines to initialize, seek, draw a line, plot a character, eject to a new frame, read graphic input, fill a polygon, reset the device to allow non-graphic output, and to wrap-up. All of these routines must be furnished, although not all of them need to actually do anything (actually, polygon fill and graphic input can be omitted completely). For example, on a device with no distinction between graphic mode and terminal mode, the reset routine may be null.

In order to make them easy to implement, all of these routines operate in the *raster* or *device coordinate* system. These are the coordinates natural to the graphic device; most devices are addressed by integral values from 0 to some specified maximum along each axis.

The device driver routines are designed to support relatively dumb graphical devices. They do not assume that the device can perform coordinate transformations (mapping user coordinates to device coordinates), or clipping. Also, the device drivers do not assume capabilities for the device to plot a set of line segments, a number of points, or even a text string. Instead, all of these operations are carried out one line or character at a time.

This normally means that some unnecessary operations are carried out on more intelligent devices that can, for example, plot text strings. However, the overhead introduced by this low-level treatment of the device is typically not important in the overall context of interactive data analysis. Readers who must use the features of a new device can read the programs that invoke the device driver operations and make their own changes.

Included next is a model for a device driver program written in C. We picked C since it is more convenient than FORTRAN for writing device drivers, since it has substantially better facilities for dealing with and transmitting ASCII characters. This model implements the basic device driver operations, although the *printf* statements contain just suggestions of what might be printed in order to control the device.

```
/* a generic S device driver, written in C */
/* should be in a .C file so that it is processed */
/* by the m4 macro processor */

#include <stdio.h>

extern float F77_COM(bgrp)[]; /* graphical parameters */
```

```
define('am','F77_COM(bgrp)[$1-1]')

int ask = 1;

F77_SUB(zparmz)(par,n)
float par[]; long *n;
{
     int i;
     for(i=1; i<=39; i++) am(i) = 0.0;

     am(20) = 10.0; /* char size in rasters */
     am(21) = 12.0;

     am(22) = 0.0;   /* x limits in rasters */
     am(23) = 399.0;

     am(24) = 0.0;   /* y limits in rasters */
     am(25) = 299.0;

     am(26) = 0.29; /* character address offset */
     am(27) = 0.85;

     /* screen is nominally 9.6 by 7.2 inches */
     am(28) = 9.6/(am(23)-am(22)); /* raster size inches */
     am(29) = am(28);       /* assumes square rasters */

     am(30) = -8000.;      /* device code number negative */

     am(31) = 1.0;   /* characters can be rotated */
     am(1) = 1.0;    /* characters are variable size */

     if(*n>=1) ask = par[0];

     printf("INIT");/* initialize graphics device */
}

F77_SUB(zseekz)(ix,iy)
long int *ix,*iy;
{
     static float line_type = -1., line_width = -1.;

     if(line_type!=am(8)){     /* set line type if changed */
          line_type=am(8);
          printf("T(%d)",(int)line_type);
          }

     if(line_width!=am(9)){    /* set line width if changed */
```

```
                     line_width=am(9);
                     printf("W(%d)",(int)line_width);
                     }

              printf("M(%d,%d)",*ix,*iy);    /* move to ix, iy */
       }


       F77_SUB(zlinez)(ix,iy)
       long int *ix,*iy;
       {
              printf("L(%d,%d)",*ix,*iy);    /* line to ix, iy */
       }


       F77_SUB(zptchz)(ich,crot)
       char *ich; float *crot;
       {
              static float size = -1., rot = -1.; /* old values */

              if(size!=am(18)){   /* set character size */
                     size=am(18);
                     printf("S(%d,%d)",(int)(size+.5),(int)(size+.5));
                     }

              if(rot != *crot){   /* set character rotation */
                     rot = *crot;
                     printf("R(%d)",(int) rot);
                     }

              printf("C(%c)",*ich);    /* print the character */
       }


       F77_SUB(zejecz)()
       {
              if(ask){
                     printf("GO? ");
                     while( getchar() != '\n');    /* ignore input */
                     }
              printf("E");   /* clear the screen */
       }


       F77_SUB(zflshz)()
       {
              fflush(stdout);
       }


       F77_SUB(zwrapz)()
       {
```

```
        printf("W");
    }
```

The initialization routine

zparmz(par,n)

is called once, before any other device driver routines are called. It has the responsibility to initialize the device and to set up certain values in the graphical parameter array. In particular, it should define the basic parameter values in the *am* array. These parameter values can usually be obtained easily from the programmer's manual for the graphic device. The only non-obvious parameters are $am(26),am(27)$, that tell where a character is addressed on the current device. For example, the values (0,0) indicates that the lower-left of the character is the addressed position, (.5,.5) that characters are centered at the current device position, etc. Sometimes the correct values of these parameters may be obtained by experimentation; try plotting a large character atop a cross and note whether it is centered.†

Certain of these parameters are somewhat rough indicators of the capabilities of the device. If the character size change or rotation parameters indicate that the device has these capabilities, it is assumed that both parameters are infinitely variable, although many devices have only a discrete number of sizes and rotations.

In *zparmz*, the argument *par* provides a list of *n* device-specific parameters. In many cases, *n* may be zero. For certain devices, these parameters can describe device characteristics such as width, height, etc. See section 3.6.4.

The basic plotting operations on the device are carried out by analogy with a pen-plotter device. The routine

zseekz(x,y)

causes a *seek* operation, i.e., the device positions itself (without drawing anything) to the coordinates (x,y). (Remember, these are integer raster coordinates).

The device driver can now either draw a line from its current position to a new position, or it can plot a character at the current position. These operations are carried out by routines

---

† The test routine "adm/test/device" under the S home directory provides a test of various device parameters, and can be useful in checking out a new device driver.

    zlinez(x,y)
    zptchz(char,rot)

The character plotting routine should interpret the argument *rot* as a rotation angle for the character, measured in degrees counter-clockwise from horizontal.

Most often, the *zseekz* routine takes responsibility for control-ling changes of color or line type if the device has those capabilities. Routine *zptchz*, in addition to changing character rotation, also should set character size according to the *cex* parameter.

Routine *zejecz* is called whenever it is necessary to advance to a new frame on the graphic device. On many devices, it will issue a message to the user such as "GO?", which allows the user time to put a new sheet of paper in a plotter or to make a copy of information on a scope. Once a reply is received to the message it clears the screen.

Some devices provide a distinction between *alphanumeric* mode and *graph* mode. Others, e.g. pen plotters, may be turned on or off to enable or disable graphics. In either case, it is important that, at the conclusion of graphic output, the device be left in a mode suitable for normal terminal interaction. This is the responsibility of routine

    zflshz

which flushes any remaining graphic commands out to the device and restores alphanumeric mode. Routine *zseekz* is always the first routine called to resume plotting, and it should take responsibility for ena-bling graph mode.

Some care should be taken to make the alphanumeric/graph mode switching reasonably efficient. In a system of graphical subrou-tines, it is not possible to know when the calling routine will do alphanumeric output to the terminal. So, to be on the safe side, the graphical subroutines return the terminal to alphanumeric mode whenever they give control back to the caller. This may mean lots of mode switching.

When the device driver is called for the final time, *zwrapz* is invoked to wrap-up operations. In many cases, nothing special is necessary for wrap-up, although pen plotters may put away the pen, etc.

Reading the source code for current device drivers is perhaps the best way to get a feel for how a new device driver should be implemented. Try looking at code in directories named, for example, "src/fun/hpgl" under the S home directory.

**3.6.2** *Portability Considerations*

Many of the older device drivers within S were written in
FORTRAN with the intention that they could be ported to other computing environments. Three routines are provided in the support
library to achieve the portability of FORTRAN-based device driver
code.

Most interactive devices are prepared to receive ASCII characters from a computer system, and to recognize special combinations of
these characters as graphic commands. Unfortunately, algorithms
written in a FORTRAN-based language have no capability to communicate some of these characters to the device. The routines

```
call zoutrz(buf,n)
call zoutwz(buf,n)
ich = i6to9z(char)
```

were designed to remedy this situation. The first two take an integer
vector *buf* and interpret its contents as coded ASCII characters. For
example, 12 is the code for the "form-feed" character. The first *n*
values in *buf* are sent to the terminal as *n* ASCII characters, with no
appended carriage return, etc. In addition, *zoutwz* waits until some
input is received from the terminal before returning (normally used
when the user is prompted for permission to clear the device and go
to the next frame.)

The function *i6to9z* translates a single character from the
machine's internal code to the ASCII code used by *zoutrz* and *zoutwz*.
It is normally used by *zptchz* to turn characters into ASCII for the output device.

By using these routines to communicate with a graphic device,
it is possible to be certain of the ASCII characters that will reach the
device, and yet avoid any knowledge of the machine's internal character set.

**3.6.3** *Control of Graphic Input*

One of the most difficult parts of writing a device driver is providing for graphic input. All graphic input is done by the user-
written routine

```
zquxyz(x,y,indic)
```

where *(x,y)* is the set of device coordinates for the selected point, and
*indic* is an indicator that can be used to show status information. One
special use of *indic* is that negative values signal end of graphic input;

positive values can be used to code pen status (up or down), current color, character used to transmit input, etc.

One of the most difficult tasks with graphic input is controlling line turnaround, full/half duplex transmission (echoed characters may interfere with the device), etc. Because they are tailored to the environment, graphic input routines are typically non-portable.

### 3.6.4 *Writing a Device Driver for S*

To create an S device driver function, use the "−d" flag on the FUNCTION utility. For example, if we acquire a new Zork-7000 graphics device and want a device function *zork*,

$ S FUNCTION −d zork zork.i

With this flag, the utility does a number of things. A directory named *zorksource* is created to hold the source files for the device driver routines (zparmz, etc.). In addition, provision is made for two new S functions: *zork* and *dev.zork*.

To create the device driver, write the device dependent routines as outlined in this section, and place them on files in directory *zorksource*.

Next, write an S Function *zork.i* as follows:

```
#Driver for zork7000 graphics terminal
FUNCTION zork( )
DEVICE_DRIVER
END
```

Once the interface routine is written, use the *MAKE* utility to create executable versions of the interface routine and the associated device driver routine.

$ S MAKE zork dev.zork

All routines in the device source directory are compiled and loaded in the *dev.zork* routine.

The interface routine can serve a more interesting purpose if *zparmz* allows optional parameters. In this case, if the interface routine returns a vector named *parms*, it will be passed on to *zparmz*. For example, if our Zork-7000 is a pen plotter with variable paper size,

```
#Driver for zork7000 pen-plotter
FUNCTION zork(
        width      /REAL,1,10./
        height     /REAL,1,8./
        )
STRUCTURE(parms/REAL,2/)
```

```
parms[1]=width
parms[2]=height
DEVICE_DRIVER
RETURN(parms)
END
```

Routine *zparmz* will be called with a vector containing the two parameters. It can use this information to set up an appropriate device coordinate system, character size, etc. This will allow users to invoke the *zork* device function with optional parameters specifying the paper width and height.

# Appendix A
# Interface and
# Algorithm Languages

This appendix presents a set of tables describing various features of the S interface and algorithm languages. There is also a semi-formal grammar describing the interface language.

## A.1 Interface and Algorithm Languages

Because the interface and algorithm language compilers are actually made up by combining tools following the philosophy of the UNIX brand operating system, the languages inherit a number of idiosyncrasies of the tools that are used to process them. In particular, the algorithm language uses the *m4* macro processor and the *ratfor* FORTRAN pre-processor. The interface language uses these two programs, along with a specialized processor of its own.

---

### Reserved Words in the Algorithm and Interface Languages
#### (*Ratfor* and *m4* reserved words)

| | | |
|---|---|---|
| break | eval | popdef |
| case | for | repeat |
| decr | if | shift* |
| define | ifdef | substr |
| defn | ifelse | switch |
| divert | incr* | syscmd |
| divnum | index* | sysval |
| dnl | len* | until |
| do | next | while |
| else | | |

* Experience has shown that these names are the most likely to conflict with user-defined variable names.

---

### Reserved Words in the Interface Language
#### (variables generated by the interface macros)

| | | |
|---|---|---|
| curkey | ls | rs |
| instr | naflag | snull |
| instra | nres | xl |
| is | nx | xu |
| lablen | ny | yl |
| logx | outstr | yu |
| logy | | |

The interface language generates a number of variable names containing the letters "qq" or beginning with "q". Beware constructing such names for STATIC variables.

| Alphabetical List of Interface Language Macros | | |
|---|---|---|
| ALLARG | INSERT | P |
| ALLOCATE | INTEGER | PDATA |
| ARG | KEYNAME | PDIM |
| ARGSTR | LDATA | PLOTARGS |
| CDATA | LENGTH | PORT |
| CHAIN | LIKE | PTSP |
| COERCE | LOGICAL | PUTSEED |
| COPY | LOGPLOT | RDATA |
| CTABLE | MAXIMIZING | RETURN |
| DATA | MINIMIZING | SETMODE |
| DATANAME | MISSING | SETUP |
| DEBUG | MODE | STATIC |
| DERIVATIVE | MODECALC | STRUCTURE |
| DEVICE_DRIVER | NAME | TEND |
| END | NAOUT | TNPER |
| ENDARGS | NARGS | TSPLOT |
| FATAL | NASET | TSTART |
| FIND | NCOL | VALUE |
| FROM | NEXTARG | VARIANCE |
| FUNCTION | NOPRINT | VPTR |
| GETSEED | NO_TS | VSET |
| IDATA | NROW | WARNING |
| INCLUDE | ONFATAL | YESPRINT |
| INITIAL | | |

| Standard Algorithm Language Macros | |
|---|---|
| ASSERT(expression) | Abort if *expression* is FALSE. |
| CASE(variable) | Branch of SWITCH statement; *variable* is value or expression. |
| CHAR | Dynamic storage allocation, character data mode. |
| DBL | Dynamic storage allocation, double precision data mode. |
| DEFAULT | Default branch of SWITCH statement; executed if no other CASEs selected. |
| FALSE | Logical FALSE. |
| FATAL(message) | Terminate with fatal error message. |
| FATAL(message) | Terminate with routine name and fatal error message. |
| INT | Dynamic storage allocation, integer data mode. |
| LGL | Dynamic storage allocation, logical data mode. |
| MESSAGE(message) | Print message. |
| NULL | Null (zero) pointer. |
| POINTER | Pointer (integer) data declaration. |
| REAL | Dynamic storage allocation, real data mode. |
| ROUTINE(name, description) | Declare S algorithm name and purpose. |
| STRING(string) | Pointer to null-terminated string on stack. |
| SUPPORT(name, description) | Declare S support routine name and purpose. |
| SWITCH(variable) | Generalized case statement; *variable* optional. |
| SYSTEM(cmd) | Execute UNIX system command. |
| TERMINATE(message) | Terminate all execution -- cause S to die. |
| TRUE | Logical TRUE. |
| TSTRING(string) | Null-terminated string. |
| WARNING(message) | Print warning message. |

| Machine-Dependent Algorithm Language Macros | |
|---|---|
| BACKSPACE | Backspace character (\b). |
| BIG | A very large real number (near maximum size). |
| CHARACTER(name, len,subscr) | Declaration of character variable. |
| DEG2RD | Multiplicative conversion factor from degrees to radians. |
| EOS | End of string character (NULL). |
| ERRORFC | Standard error file descriptor. |
| ESCAPE | Escape character (\). |
| I1MACH(i) | PORT constants (compile time). |
| INCLUDE(files) | Includes standard macros. |
| LARGEINT | A very large integer (near maximum size). |
| NA | Missing value. |
| NA(x) | Test for missing value. |
| NASET(x) | Set $x$ to missing value. |
| NBPC | Number of bits per character. |
| NCPW | Number of characters per word. |
| NDIGITS | Approximate number of decimal digits of precision in real numbers. |
| NEWLINE | New-line character (\n). |
| NOARG | Flag indicating missing argument. |
| PI | The value of pi, 3.14159.... |
| PRECISION | Small number such that 1+PRECISION != 1. |
| R1MACH(i) | PORT constants (compile time). |
| SAVE(vars) | Generates f77 save statement for variables that must retain values. |
| STDIN | Standard input file descriptor. |
| STDOUT | Standard output file descriptor. |
| TAB | Tab character (\t). |

The next 6 tables show macros defined in the interface language, in addition to the macros from INCLUDE(print,stack,struct).

| Data Structure Manipulation | |
| --- | --- |
| CDATA(n,i,j,k) | Access n[i,j,k] assuming character pointer data; (j,k) optional. |
| DATA(n,i,j,k) | Access n[i,j,k] using previously declared mode; (j,k) optional. |
| DATANAME(n) | Pointer to character actual argument name. |
| IDATA(n,i,j,k) | Access n[i,j,k] assuming integer data; (j,k) optional. |
| KEYNAME(n) | Pointer to character formal argument name. |
| LDATA(n,i,j,k) | Access n[i,j,k] assuming logical data; (j,k) optional. |
| MISSING(n) | FALSE if argument $n$ found; TRUE if default action occurred. |
| NAOUT(n) | Gets rid of NAs in $n$. |
| NCOL(n) | Number of columns of matrix. |
| NROW(n) | Number of rows of matrix. |
| P(n) | Pointer to data structure. |
| PDATA(n) | Pointer to Data entry of data structure. |
| PDIM(n) | Pointer to Dim entry of matrix or array structure. |
| PTSP(n) | Pointer to Tsp entry of time-series structure. |
| RDATA(n,i,j,k) | Access n[i,j,k] assuming real data; (j,k) optional. |
| TEND(n) | End time of time-series (real number). |
| TNPER(n) | Number of periods per cycle of time-series (real number). |
| TSTART(n) | Start time of time-series (real number). |
| VPTR(n) | Pointer to actual list of data values. |
| VSET(n) | Bring VPTR up-to-date (internal use). |

| Declarations | |
|---|---|
| CTABLE(name, string1, ...) | Set up table of character strings (see *match*). |
| INITIAL(name/value/, ...) | Generate f77 data statement. |
| INTEGER | Equivalent to INT. |
| LOGICAL | Equivalent to LGL. |
| STATIC(declaration) | Make static f77 declaration (integer, real, etc.). |

| Arguments | |
|---|---|
| ALLARG(n) | Loop setting $n$ to each unmatched argument in turn. |
| ENDARGS | End of argument list. |
| FNAME | Pointer to name of the function. |
| NARGS | Number of arguments given to function. |
| NEXTARG | End of an *ALLARG* loop. |

| Miscellaneous | |
|---|---|
| GETSEED | Read dataset *Random.seed* and set up for random number generator. |
| PUTSEED | Write current state of random number generator to dataset *Random.seed*. |
| DEBUG | Enable interactive debugging. |
| DEBUG(n) | Print data structure $n$ and enable interactive debugging. |
| DEBUG(x,n) | Print FORTRAN vector $x(n)$ and enable interactive debugging. |
| PORT | Declare intent to use PORT library routines (disable PORT error handler). |

| Returned Values | |
|---|---|
| NOPRINT | Executable statement; mark result to not be automatically printed. |
| YESPRINT | Executable statement; mark result to be automatically printed. |
| RETURN(n,m,..) | Return data structures. |
| RETURN(FILTER) | Return any data structures not picked up as arguments. |
| INSERT(n,v) | Insert data structure $v$ into structure $n$. |
| CHAIN(name,n,m,...) | Invoke function *name* on returned values. |
| ONFATAL(expression) | Executed if fatal error happens (used for cleanup, etc.). |

| Graphics | |
|---|---|
| NO_TS | Do not allow time-series axes. |
| TSPLOT | Logical, is this a time-series plot? |
| PLOTARGS(NAOK) | Pick up x,y arguments from arg list. Allow x-y structure or time-series. Are NAs allowed in x,y (default FALSE)? |
| LOGPLOT | Allows logarithmic axes, looks for argument *log=*. Sets logical variables logx and logy. |
| SETUP(xmin,xmax, ymin,ymax) | Set up axes; looks for *xlim=* ,*ylim=*. Transform data $x$ and $y$ if logs needed. Calls *beginz*. |
| PAR | Argument representing all graphical parameters given in *name=value* form. Parms set during execution of function; reset unless PAR is returned. |

| Data Structure Macros From INCLUDE(struct) | |
|---|---|
| ALLENTRY(d,p) | Loop, letting *p* range over all entries of *d*. |
| DIR | Mode signalling a directory structure. |
| DPOS(d,p) | Number of entry pointed to by *p* of structure *d*. |
| ENTRY(d,n) | Pointer to *n*th entry of structure *d*. |
| FIRSTENT(d) | Pointer to first entry in structure. |
| LACTUAL(d) | Allocated length of structure directory. |
| LASTENT(d) | Pointer to last entry in structure. |
| LENGTH(p) | Length component of entry. |
| LENTRY | Number of integers per entry. |
| LHDR | Number of entries per directory header. |
| MODE(p) | Mode component of entry. |
| NAME(p) | Name component of entry. |
| NEXTENT(p) | Pointer to next entry after *p* in structure. |
| SNAME(d) | Name of structure directory. |
| STR | Mode signalling a data structure entry. |
| STYPE(d) | Obsolete type of structure directory. |
| USED | Flag for NAME field of entry, signalling it has been used. |
| VALUE(p) | Value pointer component of entry. |

| Printing Macros<br>From INCLUDE(print) | |
|---|---|
| ABORT(list) | Encode items in *list*, print on ERRORFC, abort. |
| BUFFER | Character variable, the print buffer. |
| BUFLEN | Maximum length of print buffer. |
| BUFPOS | Current position in print buffer. |
| CLEAR | Clear the print buffer. |
| ENCODE(list) | Encode items in *list* into print buffer. |
| EPRINT(list) | Encode items in *list*, print on ERRORFC, clear buffer. |
| FPRINT(fc,list) | Encode items in *list*, print on file *fc*, clear buffer. |
| PRINT(list) | Encode items in *list*, print on STDOUT, clear buffer. |
| SKIP(fc,n) | Skip *n* lines on file *fc*. |

| List items | |
|---|---|
| "string" | Quoted string (no commas allowed). |
| C(c,w) | Character string, optional width. |
| COMMA | A comma. |
| F(r,p1,p2,p3) | Real, optional parameters width, decimals, type. |
| I(i,w) | Integer, optional width. |
| L(i,w) | Logical, optional width. |
| NASYMB | The characters NA. |
| O(i,w) | Integer in octal, optional width. |
| Q(c,w) | Quoted character string, optional width. |
| R(r,p1,p2,p3) | Real, optional parameters width, decimals, type. |
| S(c,p1,p2,p3) | Character string, optional parameters width, leng, quotes. |
| SHARP | A sharp (#) character. |
| SP(n) | Space *n* characters. |
| T(pos) | Tab (move) to position *pos*. |
| TABCHAR | A tab character. |
| V(p,mode,p1,p2,p3) | Pointer *p* to data value of mode *mode*, optional parameters. |

| Reading Macros<br>From INCLUDE(read) | |
|---|---|
| DECODE(list) | Decode items in *list* from input buffer. |
| EOF | End of File. |
| FIELD_ERROR | Logical, indicates that a field had an error |
| FREAD(fc,list) | Read *fd*, decode items in *list*. |
| GETLINE(fd) | Read line from file *fd*. |
| INBUF | Input buffer. |
| INFILE | Current input file descriptor. |
| INLEN | Maximum length of input buffer. |
| INPOS | Current position in input buffer. |
| READ(fc,list) | Read STDIN, decode items in *list*. |
| UNGETC | Give back character to buffer. |
| UNGET_FIELD | Give back field to buffer. |

| List items | |
|---|---|
| C(c,len) | Character string of given length. |
| CP(p) | Pointer to character string on stack. |
| GETC(c) | Next character. |
| I(i) | Integer. |
| L(l) | Logical value. |
| N(i) | Non-negative integer. |
| NP(p) | Pointer to name on stack (unquoted string). |
| O(i) | Octal integer. |
| R(r) | Real. |
| S(c,len) | Character string of given length. |
| V(p,mode) | Pointer *p* to data item of mode *mode*. |

| Graphics Macros<br>From INCLUDE(graphics) | |
| --- | --- |
| am | Graphical parameter array (used only by QUERY, SPECIFY). |
| AXIS(side,umin,umax) | Set up axis ranging from *umin* to *umax*. |
| BOTTOM | Side of plot (==1). |
| LEFT | Side of plot (==2). |
| QUERY( par(values), ...) | Query graphical parameters. |
| RIGHT | Side of plot (==4). |
| SPECIFY( par(values), ...) | Specify graphical parameters. |
| TOP | Side of plot (==3). |

| Stack Allocation Macros<br>From INCLUDE(stack) | |
| --- | --- |
| cs | Stack as character. |
| ds | Stack as double precision. |
| is | Stack as integer. |
| ls | Stack as logical. |
| rs | Stack as real. |
| TEXT(p) | Character string located at pointer *p*. |

---

Strange Happenings; Warts

---

m4 doesn't recognize quoted strings, thus
      macro("this is a string, containing a comma")
is interpreted as a macro with two arguments.

m4 macros must have argument lists appear immediately after the macro name; there can be no intervening white space; thus
      macro(arg)
      macro (arg)
are very different in effect. The first gives one argument to the macro; the second invokes the macro with no arguments.

The characters grave (`) and apostrophe (') are used as quotes for m4, and should not be used for other purposes.

Because of limitations of the interface language, arguments to the RETURN macro should not have trailing blanks.

---

## A.2 Grammar for Interface Language

The interface langauge is a super-set of the algorithm language, which in turn is a superset of RATFOR. The following grammar describes the two extensions in terms, primarily, of the definition of a *statement* in the three languages. Because the interface language is actually processed by a set of tools, not by a parser based on *yacc*, the precise limits of the language are not compactly describable by a *yacc* grammar. However, the grammar below, along with some explanatory notes on the main types of statement, will clarify what is legal in the language.

| | | |
|---|---|---|
| function | : | "FUNCTION" name "(" list ")" body "END" |
| body | : | statement |
| | \| | body separator statement |
| statement | : | interface_stmt |
| | \| | algorithm_stmt |

```
separator      :  ";" | "\n"

interface_stmt:  "STRUCTURE(" list ")"
             |   "RETURN(" ret_list ")"
             |   "RETURN"
             |   "CHAIN(" ret_list ")"
             |   "INSERT(" ret_list ")"
             |   "COERCE(" name "," mode_type ")"
             |   arg_stmt
             |   graph_stmt
             |   debug_stmt
             |   special_decl
             |   special_control

arg_stmt       :  "ARG(" list ")"
             |   "ALLARG(" name ")"
             |   "ALLARG(" name "," NOKEY ")"
             |   "NEXTARG"
             |   "ARGSTR(" name ")"
             |   "ARGSTR"

graph_stmt     :  "PLOTARGS ( NAOK )"
             |   "PLOTARGS ( NAFATAL )"
             |   "PLOTARGS"
             |   "LOGPLOT"
             |   "SETUP"
             |   "SETUP (" r_expr "," r_expr ","
                     r_expr "," r_expr ")"
             |   "QUERY (" pars ")"
             |   "SPECIFY (" pars ")"
             |   "AXIS (" i_expr "," r_expr "," r_expr ")"
             |   "DEVICE_DRIVER (" name ")"

debug_stmt     :  "DEBUG"
             |   "DEBUG ( ON )"
             |   "DEBUG (" namelist ")"

list           :  /* empty */
             |   item
             |   list "," item

item           :  name
             |   name "/" decls "/"
```

| namelist | : | name |
| | | namelist "," name |

| decls | : | mode_type |
| | | mode_type "," values |

| mode_type | : | mode |
| | | type |
| | | mode "," type |

| mode | : | "LGL" | "INT" | "REAL" | "CHAR" | "ANY" |
| | | "MODECALC(" i_expr ")" |

| type | : | "VECTOR" | "MATRIX" | "ARRAY" | "TS" |
| | | "STR" | "NULL" |

| special_decl | : | "STATIC (" anything ")" |
| | | "INITIAL (" anything ")" |
| | | "NOPRINT" |
| | | "YESPRINT" |

| ret_list | : | ret_item |
| | | retlist "," ret_item |

| ret_item | : | name |
| | | name = name |
| | | = name |

| opt_stmt | : | "FIND(" name ")" |
| | | "MINIMIZING(" name "," algorithm_stmt ")" |
| | | "MAXIMIZING(" name "," algorithm_stmt ")" |
| | | "DERIVATIVE(" name "," algorithm_stmt ")" |

The nonterminal *algorithm_stmt* stands for any statement acceptable to RATFOR.

Special forms of the nonterminal *item* are recognized in the FUNCTION and ARG statements; namely, the reserved names PAR, FITTING, and &. Similarly, the nonterminal *ret_item* has the special forms FILTER, PAR, VARIANCE, and DERIVATIVE in the RETURN statement. In the CHAIN and INSERT statements, the first *ret_item* must be a name (respectively, of a function and of a structure).

The nonterminals *i_expr* and *r_expr* are integer-valued and real-valued expressions in the algorithm language; *name* is a name token in the algorithm language. The nonterminal *special_control* refers to the SWITCH MODE statement described in section 1.2.5; it has essentially the syntax of a *switch* statement in ratfor, with different tokens SWITCH, CASE, and DEFAULT, and with some semantic restrictions on the case constants.

# Appendix B
# Definition of the S Language
# and Data Structures

This appendix describes the S language and system more formally. It is the place to find precise definitions of what is legal (syntax), what is the effect of expressions (semantics), and the full extent of the data structures and other system components. In addition, a number of specific features and details are introduced or elaborated.

## B.1 The Language: Syntax

The process of evaluating user's expressions in S proceeds in two phases: first, the expression as typed is *parsed* to produce a data structure; then this data structure is *interpreted* to evaluate the individual function calls in the expression. This subsection describes the parsing process, which in turn is done in two steps. User input goes to a *lexical analyzer*, a program which breaks the input into *tokens* (names, numbers, character strings, etc.). These tokens are passed to a *parser* which interprets the form of the expression as a whole. In the case of S, the parser operates on the basis of a set of *syntax rules*, specifying the form of all legal expressions. In fact, the rules are relatively compact and straightforward (this is one reason for the flexibility of the language). We will present in this section the *complete* syntax for S, with only a few alterations to enhance readability.

Each syntax rule specifies a *syntax type* and then enumerates all the possible forms which are recognized as producing this type. These forms are defined by concatenating three kinds of symbols: other syntactic types, literals (characters to be taken as themselves), and *token types*.

**Lexical Analysis.** A token type is some class of character strings which is recognized by a lexical analyzer, prior to parsing the S expression. The token types recognized in S are as follows.

INT           an integer literal;

REAL         a real (decimal fraction or exponential) literal.

NAME       a string of characters, consisting of letters, digits, and ".", with the first character a letter. Depending on implementation, there may be a limit to the length of the name (in many implementations of UNIX, for example, names must be unique in 14 characters).

STRING      any string of characters enclosed in matching, unescaped double or single quotes.

OP            a (binary) operator, one of the arithmetic or logical operators, ":", or a special operator consisting of "%" followed by any single character.

UNARY      one of the permissible unary operators ("+", "−", and "!").

MACRO      "?" immediately preceding a name indicates a call to an S macro (see note 4 below).

**Notes:**

1 — with the exception of STRING, no token can have internal white space, and, specifically, real numbers in exponent notation cannot have internal blanks;

2 — tokens of type NAME are approximately the legal names of functions and data sets (with the respective exceptions of the existence of operators and the use of *prefix* — both of which are discussed below);

3 — the type OP is in fact subdivided, but this is only relevant for questions of precedence;

4 — the MACRO token will not be recognized inside a string or in a comment. When it is recognized, expression input is continued until a possible complete expression is obtained. The parse is then terminated and the text of the expression is given to the *macro* function. This function (if it does not encounter an error) expands expands all macro calls, and writes the expanded expression to a file. A hidden invocation of the *source* function then brings this file back in for parsing.

5 — Comments in an S expression are recognized by the literal "#". The lexical analyzer then discards this character and the rest of

the line. Thus, comments are not processed by the parser.

When the lexical analyzer finds a token, it returns both the *type* (a code for one of the types above), and also a token *value*, when appropriate, giving the actual text or data value of the token. Thus, "3.14159" is a token of type REAL and has the obvious value as a real data item. In the syntax rules of this section, it is the token type which appears in the definitions. When we talk about the semantics of the expressions, the token value will be used.

**Syntax Rules.** In the presentation of the S syntax, the syntactic type appears on the left of the rule, followed by "∶" and the first possible form. Alternative forms are separated by a vertical bar character, "¦". For clarity of presentation, names for syntactic types are lower case, names for token types are upper case, and literals are enclosed in quotes. Syntax rules are terminated by a semi-colon. Comments in the rules go from "#" to the end of the line.

The key syntax type in S is *expr*, the general S expression. Here is the rule which defines all legal expressions.

```
expr      :   NAME "(" arg.list ")"                    #function call
          |   expr OP expr                             #binary operator
          |   UNARY expr                               #unary operator
          |   sub.expr "[" arg.list "]"                #subset
          |   asn.expr "←" expr                        #assign or replace
          |   expr "→" asn.expr                        #same, to the right
          |   INT | REAL | STRING                      #literals
          |   sub.expr                                 #can be subset
          |   asn.expr                                 #can receive assignments
          |   control                                  #iteration, conditional, etc.
```

These, with the exception of *control*, are the expressions encountered in most ordinary use of S (i.e., outside of macros). Clearly, they form a compact syntax, although very general expressions may be built out of the forms. The syntactic forms appearing on the right of the rules (in addition to *expr* used recursively) now must be defined. The syntactic type *asn.expr* is the class of things which can receive assignments: simple assignments or replacements of components and subsets.

```
asn.expr   :   com.name                               #name or component
           |   com.name "[" arg.list "]"              #subset
```

and *com.name* is a name, optionally followed by any number of named or numbered components:

```
com.name    :   NAME
            |   com.name "$" NAME
            |   com.name "$" "[" expr "]"
```

The syntactic type *sub.expr* is the class of things of which subsets can be taken:

```
sub.expr    :   "(" expr ")"                        #anything in parens
            |   NAME "(" arg.list ")"               # function call
            |   sub.expr "$" NAME                   #component
            |   sub.expr "$" "[" expr "]"           #numbered component
```

However, the most common form of *sub.expr* is just NAME, which is not included in the rules above!  This case is covered by the last line of the definition of *asn.expr*.  To include the same form in the rule for *sub.expr* would be ambiguous.

The syntactic rule *control* provides conditional expressions, iteration and compound expressions:

```
control     :   "if" "(" expr ")" expr
            |   "if" "(" expr ")" expr "else" expr
            |   "for" "(" NAME "in" expr ")" expr          # iteration
            |   "while" "(" expr ")" expr
            |   "repeat" expr
            |   "break" | "next"                           #loop control
            |   "{" exp.list "}"                           #compound
```

The remaining syntactic forms *exp.list* and *arg.list* are lists of expressions separated by ";" and of arguments separated by ",".

```
exp.list    :   expr
            |   exp.list ";" expr                          #see note in §B.1.2

arg.list    :   arg
            |   arg.list "," arg
```

While *expr* has been defined, *arg* has not.  A function argument can be empty, or an expression, or a "name=" followed by an empty or an expression.

```
arg     :   # empty
        |   expr
        |   NAME "="                                       #empty
        |   NAME "=" expr
```

This completes the essential rules. Now for some special cases, explanations and nits.

### B.1.1 *Function Calls and Commands*

As a concession to those who like a command-language appearance rather than a functional notation, function calls in the top-level S expression can use blanks, rather than parentheses, to separate the argument list from the function name. This is done by defining a syntactic type *command:*

command  : expr                                  *#as before*
         |  NAME BLANK arglist

The semantics of the second form is the same as that of the function call in the first line of the *expr* definition.

### B.1.2 *Compound Expressions*

The formal syntax implies that S insists on a semi-colon between expressions within a compound expression. Actually, the language is more liberal. A new line can be used to separate expressions, if the end of the line could be the end of an expression. This is accomplished outside of the parse, by the mechanism of returning ";" from the lexical analyzer when a new line occurs in this circumstance.

### B.1.3 *Continuation*

An interactive language interpreter must decide when an expression is complete, in order to execute it. (By contrast, a compiler can read a complete *program* before beginning the parse.) In S, the first legally complete expression will be parsed and executed. This becomes an issue in practice only when the user wants to type a new line, continuing the expression onto the next line. Continuation can be forced if either of the following holds:

1 — the last token is not one of those which could legally end an expression;
2 — there is an unbalanced "(" or "[".

The tokens which can end an expression are any literal (INT, REAL, NAME or STRING), ")", "]", or "}". To force continuation, manage to end the line with anything else. A special extension of the continuation rule is applied, however, so that conditional or iterative expressions can be written with the test on one line and the executed expression on the next. For example, after seeing

if(x>0)

the parser will expect a continuation, even though this would be a complete expression if "if" were a function name. The mechanism is to recognize the special tokens "for", "if" etc., and not signal parenthesis balance on reaching the ")". In complete conditional expressions, however, it is essential to get the "else" token in on the same line; otherwise a simple "if" expression is evaluated:

> if( min(x)>0 )log(x) else
>             abs(resid)

but not

> if( min(x)>0 )log(x)
> else abs(resid)

### B.1.4 *Reserved words*

Notice that several of the literal items in the syntax rules are also legal NAME tokens: "if", "else", "repeat", "while", "for", "in", "next", and "break". These can not be used as names; the lexical analyzer will recognize them first in their special sense. A syntax error will almost certainly follow.

## B.2 Semantics

This section describes what happens after an expression is parsed: how the S executive scans and evaluates the expression. An expression parsed according to the rules of section B.1 becomes a hierarchical data structure. For example, a function call becomes a structure whose components are the structures resulting from parsing the arguments to the function. The S executive scans the structure and executes all the functions invoked, proceeding from the "lowest" level up. As each substructure is evaluated, the corresponding result (e.g., the data structure returned from a function call) replaces the parsed subexpression. When the entire structure has been scanned and evaluated, the parsed expression is replaced by its value.

The remainder of this section discusses the semantics of specific parts of the S language.

### B.2.1 *Functions and Operators*

The call to an S function in an expression can, naturally, only be evaluated after all the subexpressions appearing in the argument list of that call have been evaluated. The subexpressions in a given argument list are evaluated from left to right, a distinction which is relevant only when the subexpressions have side effects (e.g., assignment):

> plot(xvar ← read(xfile), reg(xvar,y)$resid)

reads the file and assigns *xvar* before evaluating the regression. Expressions relying on the left-to-right evaluation are bad programming style, as their interpretation is usually obscure.

   After the arguments are evaluated, the function call is a structure, whose i-th component is the data structure representing the evaluation of the i-th argument in the argument list. If the argument appeared in the "name=" form, then the component has a name which is the value of the corresponding NAME token. Otherwise the component has a null name. Notice that the NAME token in this case turns into the name of a structure component: it has nothing to do with the name of either a dataset or a function.

   After the arguments are evaluated, the executive searches for a function with the desired name. The default search list is the set of functions included in the local S implementation. If one or more calls to the *chapter* function have occurred, each named chapter is placed on the search list (by default, on the front of the list). Since the search terminates as soon as a function of the given name has been found, one can redefine standard functions by putting a function with the same name on a user's chapter.

   Once the function name has been found, the S executive transfers control to the corresponding function, giving it the *argument structure*; i.e., the data structure representing the evaluated form of the actual arguments to the function. The processing of the function arguments and all further interpretation is the job of the individual function. The S executive has no knowledge of how many or what kinds of arguments are required by any functions.

   When the function has completed execution, it returns its result as a data structure to the executive, provided execution terminated normally. The result replaces the call to the function in the parsed expression and evaluation continues. If an error or user interrupt occurred during the function execution, the execution of the entire current expression is stopped, the current expression may be written to the file "sdump" for use by the *edit* function, and the executive attempts to get the next expression from the user.

   *Argument processing* is entirely the choice of the individual function. Since argument processing is interpretive, anything is possible and the actual variety is great. However, there is a simple form of argument matching that is used by most functions. Details are given in section 1.3; we give here the rules as they affect use of the S language. The function has a set of formal arguments, with the names and the order implied by the detailed documentation for the specific function. For example, the *regress* function has full calling sequence in the following form:

regress(x, y, wt, int, print, names, q)

The formal arguments are matched to the argument structure: for each formal argument, in order, a partial-match algorithm is applied. This scans each component of the argument structure. If the name of the component matches the name of the formal argument exactly, the two are matched, the search is terminated and the component is marked USED (so it cannot match any other arguments). If no exact match occurs in the complete argument structure, the argument may be matched in two other ways:

1 — there is exactly one *partial match* of names, where currently this means that the component name is a leading substring of the formal name;

2 — there is no partial match, but a component has been found with a null name; i.e., not specified in the *name=value* form and not USED. The *first* such component is matched.

In either of the cases, the component is matched to the formal argument and is marked USED. In all other circumstances, the formal argument is marked MISSING. An argument may also be made explicitly MISSING by an empty actual argument, either positionally or in the *name=* form.

After the matching process, the S function will try to *coerce* the actual argument to the data type and mode desired, if the argument is present, or will take the chosen default action, if it is MISSING. If the argument is MISSING and not optional, or if the coerce fails, an error is generated and the expression is aborted.

When all the formal arguments have been processed, the argument structure is scanned for any remaining (non-USED) components. If there are any, an error is generated. Notice that this usually ensures that misspelled argument names will be detected, even if the corresponding argument was optional. (The partial match could, but currently does not, attempt to correct misspellings and/or resolve multiple partial matches.)

Here are two examples of calls to the function *regress(x,y,wt,int,names,q)*:

```
regress( cbind(age,iq,height), income )
regress( xmat, abs(resid), y=yvar, in=FALSE )
```

In the first example, the function *cbind* is evaluated first; the data structure representing the result is an (unnamed) component of the arguments to *regress*. There will be no exact or partial match for "x"; the result of *cbind* will match positionally. Similarly, *income* will match "y" positionally. All remaining arguments will be MISSING. In the second example, "x" will again be matched positionally, this time to the data structure resulting from getting *xmat* from a data directory. However, "y" is now matched by name; it does not match *abs(resid)* which will, at the next stage, match "wt" positionally. The argument named "in" matched "int" partially; since there is no exact match, this succeeds.

From this standard form of argument processing, there are many variations. Some of the more important are as follows:

- A number of functions (e.g., *c*, *print*, *save*, and *rm*) take arbitrarily many arguments. In this case, the actual arguments may still be named, but in general there will be no name matching (with the exception below). The function has access to the argument names, which it can use if appropriate (see *save*, e.g.).
- Arguments which appear in the form *name=* in the documentation may appear only by name and will never match an unnamed actual argument. Such arguments allow for special flags to functions, like *print*, that take arbitrarily many arguments.
- The test for extra or misspelled arguments is suppressed in some functions. The most common reason is that the function will later *chain* to another function, passing on any unused arguments. For example, most high-level graphics functions chain to *title* to process the arguments *main*, *sub*, *xlab*, and *ylab*. Misspelled arguments will be detected, but not by the function originally called.

To repeat, specific functions can interpret arguments as they wish; the examples given are just the most common.

**Operators.** Semantically, there is no distinction between operators and general functions in S. Syntactically, operators cannot have named or MISSING arguments, and the rules for names of functions and names of operators are mutually exclusive, but these distinctions have no effect on the way the corresponding functions are executed.

## B.2.2 *Side Effects: Database Changes and Parameters*

The semantic effect of most S functions is as described above. The substructure representing the call to the function is interpreted, and the result replaces it in the structure representing the entire expression. There are just two ways in which functions can have additional semantic effect: by altering the contents of one of the S data directories or by changing the value of some of the internal S parameters.

Database contents may be changed by functions which create or remove datasets (*assign, save, define, restore* and *rm*) or which modify existing datasets (*edit, medit*). The function *assign* corresponds to all forms of assignment operator: "<−", underline and "−>". The *asn.expr* syntactic type will be turned into a sequence of arguments to *assign* which are either character strings, corresponding to names of datasets or components, or integers providing component numbers. The syntax of the assignment expressions restrict the dataset and component names (see the documentation for *assign*). The function *save* is similar to assignment, except that it takes multiple datasets and does not replace components or subsets of existing datasets. For each argument, *save* will use the name, if any, of the corresponding component of the argument structure as the name of the saved dataset. If there is no specified name, the argument data structure must itself have a name. The arguments given without a name can only be datasets from other data directories. Thus, *save(newx=sqrt(x))* and *save(x)* will work, but *save(sqrt(x))* will not.

The function *define* chains to *save* to store the macro definition. The macro name, extracted from the definition for each macro, becomes the dataset name of each of the macros, with "$mac." prepended. The function *edit* either chains to *assign* or creates a hidden call to *source*, if called without argument. The function *medit*, on the other hand, chains to *define*, so that references to macro arguments by name in the edited text can be converted.

Parameters internal to the S system may be set by the *options* function, by any of the functions which alter search paths for functions or data directories (i.e., *chapter, attach,* and *detach*), or by *par*, which sets graphical parameters. All such parameter settings hold until reset or until the user quits from the session. They are initialized to default values when the user logs into S. Graphical parameters are also reset to default values when the user specifies a new graphics device. Thus

```
hp7470
par(pch="+")
...
hp7470
```

leaves the plotting-character parameter, *pch*, with its default value of
"*".

### B.2.3 *Compound Expressions*

A compound expression, consisting of subexpressions separated
by semicolons or new lines and enclosed in braces, is parsed into a
structure similar to an argument list. The components of the structure
are the parsed subexpressions.

The important aspects of compound expressions (particularly
for understanding conditional and iterative expressions) involve the
order of calculations in evaluating the expression and the rules for the
value returned. With the exception of *break* and *next*, the subexpressions are evaluated in order. When a subexpression has been
evaluated, its value becomes the *current value* for the compound
expression. However, it does not replace the parsed form of the
subexpression in the compound expression (since the expression may
be iterated). The final value of the compound expression is its current
value at the time when the S executive finishes execution of the
expression. This may occur from encountering the closing "}" of an
uniterated compound expression, from execution of a *break* or *next* in
the compound expression, or from the iteration tests of *for* and *while*
expressions. In interactive processing, user interrupts can also be used
to terminate execution of a compound expression. Three forms of
subexpression do *not* alter the current value of the compound expression: *break, next,* and a simple *if* expression (no *else* expression) whose
test returns FALSE.

A frequent confusion in practice is the assumption that subexpressions in compound expressions are automatically printed. They
are not, because the automatic printing mechanism only operates at
the recognition of the *command* syntactic type; i.e., at the highest
expression level. To force printing of subexpressions, these must be
passed to the *print* function as arguments:

```
for(i in list())
    print(i,get(i))
```

prints the names and values of all datasets on the work directory.

**B.2.4** *Conditional Expressions; Iterative Expressions*

The general conditional expression

"if" "(" expr$_1$ ")" expr$_2$ "else" expr$_3$

is evaluated as follows: *expr$_1$* is evaluated and coerced to a logical vector. The first data value in the result is tested. If it is TRUE, then *expr$_2$* is evaluated and its value becomes the value of the conditional expression. If the tested value is FALSE, and no *else* expression was given, the value of the conditional expression is undefined. Otherwise, *expr$_3$* is evaluated and its value becomes the value of the conditional expression.

Of the three types of iterated expressions, the evaluation of the the expression

"repeat" expr

is a model for the others. The repeated expression is evaluated repeatedly until the occurrence of a *break* in *expr* terminates iteration, at which time the value of the *repeat* expression is the current value of *expr*. Note that a *break* expression causes a jump out of the nearest surrounding compound expression. Similarly, a *next* expression is equivalent to a jump to the point just after the last subexpression in the nearest surrounding compound expression.

Evaluation of the expression

"while" "(" expr$_1$ ")" expr$_2$

is identical to a *repeat* expression, except that before each evaluation of *expr$_2$*, the parsed form of *expr$_1$* is evaluated, coerced to LOGICAL, and the first data value of the result is tested. If it is TRUE, *expr$_2$* is evaluated; otherwise, the iteration terminates and the current value of *expr$_2$* (if any) becomes the value of the *while* expression. In evaluating

"for" "(" NAME "in" expr$_1$ ")" expr$_2$

*expr$_1$* is first evaluated and coerced to a vector of any mode. Then the value of the token NAME is assigned to an internal dataset which is a vector of length 1 and the same mode as *expr$_1$*. The data value of this internal dataset is set to the first element of *expr$_1$* before the first evaluation of *expr$_2$*, to the second element before the second evaluation, etc. Evaluation of *expr$_2$* continues until there are no more elements in *expr$_1$* or until a *break* occurs.

Notice that the dataset used is *internal* (see the discussion of names and datasets below), rather than the result of an assignment. This means that the *for* loop has no side effect on any data directory. After execution of the expression terminates the internal dataset is removed. During the evaluation of the compound expression, the internal dataset hides any dataset with the same name on one of the data directories. It is even possible to re-use the same internal name in several nested loops (which could occur if the loops were set up in different contexts, e.g. inside macros).

## B.3 Data Structures

The general S data structure can be defined recursively as follows. A data structure is a list of the form:

( name mode length $value_1$ $value_2$ ... )

where:

*name*  is a character string, in quotes, defining the name of the structure;

*mode*  is one of the vector modes (LOGICAL, INTEGER, REAL, CHARACTER) or else is of the form

STRUCTURE type

indicating a hierarchical structure ( *type* is an obsolete integer value which once identified particular forms of hierarchical data structures, e.g., arrays, time-series.) In the *dget* and *dput* functions, modes are abbreviated to single characters ("L", "I", etc.).

*length*  is the number of values (for a vector) or components (for a structure) and it is understood that there are exactly *length* occurrences of $value_i$. In the case of mode STRUCTURE, the individual values will themselves be lists of this form (vectors or further structures).

For example, the matrix defined by

> x←matrix(1:12,3,4)

would have the form

```
( "x"  S 21 2
  ( "Dim"  I 2   3   4 )
  ( "Data" I 12   1  2  3  4  5  6  7  8  9  10  11  12 )
)
```

Data may be written to a file in this form by the function *dput*. Conversely, a file containing a data structure in this form can be read into S by the function *dget*. The related functions *dump* and *restore* write and read lists of datasets. In addition the *extract* macro and utility function extract datasets from data files consisting of identically formatted records. These functions provide conversions between internal S data structures or user files and a general list representation of data structures. In particular, notice that this list form of representation for data structures is both portable and completely general. All possible S data structures can be represented in this form, without loss of information.

The list form shown here is the *external* representation of S data structures. There is also an *internal* representation, solely for reasons of efficiency. The S user should never need to know about the internal representation. Essentially, internal representation is used instead of characters to represent logical, integer and real data items, and some pointers are used to find data values. It is important, however, to understand how the different external and internal forms relate to S functions that manipulate data:

- the functions *read* and *write* transform between data items on files, without any structural information, and vectors in S;
- the functions *dget* and *dput* transform between data structures on files in fully general external form, and the corresponding structure in an S expression;
- the *get* and *assign* functions transform between datasets on an S data directory (in internal form) and the corresponding data structure in an S expression.

In most cases, data which was generated outside S will be accessed either through the *read* function or through the *extract* macro or utility function. Transmission of data, particularly large numbers of datasets, between S data directories on different machines will be accomplished by *dget, dput, dump,* and *restore.*

The relationships among files of various forms, S datasets and functions that convert from one form to another is shown symbolically in Figure 1.

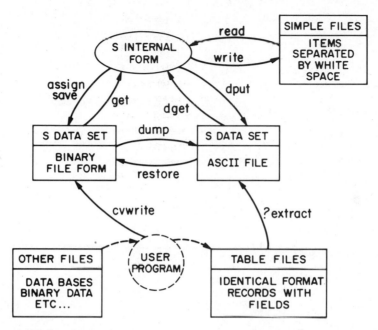

**Figure B.1.** The diagram shows the relationships among S data directories, external files, and the internal form of data structures. Arrows are marked with the names of functions that convert from one form to another.

# Appendix C
# Documentation for S Utilities

This appendix gives documentation for various S utilities that are useful when writing new S functions. Also described is the organization of the S source code.

| C | see PROGRAM | C |

| CHAPTER | Maintain a Chapter of S Functions | CHAPTER |

USAGE:

> !S CHAPTER   # UNIX command
> !S FUNCTION  [options] name files ...   # UNIX command
> !S MAKE name   # UNIX command

A user's chapter of S functions is initialized (once) under the current UNIX directory by executing the command **CHAPTER**. Each time a new S function is to be written, the **FUNCTION** command is used (once). Whenever the function has been changed (either the interface routine or any of the other files on which the function depends), the **MAKE** command is used.

ARGUMENTS:

**options**:   options controlling the interface to the function: a concatenation of any of the following: −**g** for a graphics function; −**r** for a function which needs to read the system common blocks (for example, because it does printing on OUTFC or generates random numbers) −**w** for a function which modifies the system parameters, and so must write the system common blocks (strongly discouraged for user functions), and −**d** for a device-driver function.

**name**:   the name of the S function. There should be a corresponding interface routine on a file **name.i** in the current directory.

**files**:   names of files on which the function depends (in addition to standard S libraries), including the interface routine file **name.i**, source files (which must end in ".r", ".f", ".C", or ".c") or library files. The **MAKE** command will try to keep up-to-date with source files, and will simply include all other files in the command that loads the function.

The **CHAPTER** utility creates subdirectories named "x" and ".help" in the current directory, as well as a file "Smakefile". Subdirectory "x" will contain the executable versions of any functions created in the chapter; ".help" contains the detailed documentation for each function in the chapter. The **FUNCTION** utility updates file

"Smakefile", so that it contains file names and commands for creating functions in the chapter.

Whenever **MAKE** is executed, it follows the "Smakefile" rules to create the executable file in subdirectory "x". In the process, it also creates object files, whose names end in ".x" and ".o", from the source files. These object files can be removed as soon as a satisfactory version of the function has been created. The file "i.**name**.C" is an intermediate file and can also be removed safely.

It is the user's responsibility to keep up-to-date any libraries, object files, etc. that are not given to **FUNCTION** as source files. If for some reason, the function must be reloaded even though none of the files on which it depends has been modified, precede the **MAKE** command by the UNIX command **rm name**.

If the set of files on which the function depends changes; e.g., the function now depends on a new ".r" file, the correct FUNCTION command should be given. This is the only time that FUNCTION must be re-invoked.

| **EDITDOC** | see PROMPT | **EDITDOC** |
|---|---|---|

| **FORTRAN** | see PROGRAM | **FORTRAN** |
|---|---|---|

| **FUNCTION** | see CHAPTER | **FUNCTION** |
|---|---|---|

| **MAKE** | see CHAPTER | **MAKE** |
|---|---|---|

| **NEWDOC** | see PROMPT | **NEWDOC** |
|---|---|---|

| PRINTDOC | see PROMPT | PRINTDOC |

| PROGRAM | Stand-alone Programs | **PROGRAM** |

To create stand-alone programs which use the S libraries of algorithms and/or are written with the S facilities use the command:
**!S PROGRAM [options] name files ...**   # UNIX command
**!S MAKE name**   # UNIX command

ARGUMENTS:

**options**:   optional flag, where the value **−g** indicates a graphics program that should search the graphics libraries, and **−d** indicates a device driver.

**name**:   name of the stand-alone program to be produced.

**files**:   names of files on which the program depends (in addition to standard S algorithm libraries), typically source files (which must end in ".r", ".f", ".C" or ".c") or library files.

The **PROGRAM** command is used once to specify the name of the stand-alone program and the names of source files of programs (including a main program) needed to produce **name**.

The **MAKE** command will ensure that the executable program is loaded with up-to-date object versions of the files on which it depends.

| PROMPT | Utilities for Documenting S Functions | **PROMPT** |

USAGE:

**!S PROMPT file.i ...**   # UNIX command
**!S NEWDOC file.d ...**   # UNIX command
**!S EDITDOC [−f] name ...**   # UNIX command
**!S PRINTDOC [−options][name ...]**   # UNIX command

These routines provide mechanisms to help maintain documentation for chapters of user-written functions, macros or datasets.

The **PROMPT** utility takes a list of files which contain interface routines (these files have names ending in ".i"). For each "**file**.i", **PROMPT** produces an outline of documentation on "**file**.d". The system macro **?prompt** provides a similar facility for constructing the outline of documentation for datasets and macros.

Each of the resulting ".d" files should be edited with a text editor. Within each file are a number of lines which include the character "˜" and a brief description of what should be documented at that point. These files use a number of **documentation macros** to control the formatting of the printed documentation. For example, the lines beginning ".AG" introduce function arguments; ".RC" returned components; ".FN" the names of functions documented in this file; ".TL" introduces the title; ".CS" introduces the calling sequence; ".EX" introduces lines of examples. The documentation macro ".PP" is used to start a paragraph, and results in a blank line in the printed documentation. ".IP topic" creates an indented paragraph that is labelled by **topic**. These last two macros can be used to break the documentation into appropriate paragraphs.

Once the documentation outline files have been edited, they should be installed into the chapter documentation directory by means of the **NEWDOC** utility. **NEWDOC** will automatically install macro and dataset documentation into the data directory. Once this installation is done, the documentation will be accessible via the **help** function as long as the chapter or data directory are being searched.

Once the documentation is installed by **NEWDOC**, the ".d" file can be thrown away. When it is necessary to change the documentation after it has been installed, the utility **EDITDOC** can be used. When given a list of documentation names, **EDITDOC** looks through the data directory and chapter documentation for the named documents. The documentation is then extracted and placed on file **name.d**. This file can be edited, and then reinstalled by means of **NEWDOC**. If a function has the same name as a macro or dataset, **EDITDOC** can be told to retrieve the function documentation by means of the optional "−f" flag.

The utility **PRINTDOC** is used to construct a neatly-formatted listing of documentation. By default, all macro and dataset documentation is printed on the user's terminal. If a list of names is

given, the documentation corresponding to those names is printed.  The **options** list is a set of characters preceded by a dash, and
controls the choice of output device and the listing of function
documentation.  If **options** contains the character "f", documentation for functions (rather than macros and datasets) will be printed.  The character "t" indicates that **troff** should be used to produce final documentation on the phototypesetter.  The character
"o" indicates that documentation is to be done on an offline
printer, not the terminal.

EXAMPLES:

**S PROMPT myfun.i**
**... edit the file myfun.d ...**
**S NEWDOC myfun.d**
**S PRINTDOC −ft**     # typesets all function documentation

---

| **RATFOR** | see **PROGRAM** | **RATFOR** |
|---|---|---|

---

| **System** | S System File and Source Code Organization | **System** |
|---|---|---|

S organizes its system files for source code, libraries, utilities and
functions by exploiting the hierarchical file structure in the UNIX
operating system.  The various directories to be described each
contain files related to specialized parts of S.

Source code for interface routines resides in files with names ending in ".i", and algorithm source file names will end ".r".  Include
files end with ".m", C programs that use the S **m4** macros end in
".C", and standard C programs end in ".c".

On any installation, all the S files will be under the S home directory.  The S home directory is divided into subdirectories: adm,
cmd, doc, graph, newfun, s, and src.

adm:         The portion of the file system used by the S administrator.

adm/bkup:    Back-up versions of S functions from s/x.

adm/cmd:         Administrator utilities, used in compiling S, etc.

adm/test:   Files which can be used to test the installation of S; test runs and the expected output.

cmd:         The executable commands (utilities) in S. The command **S name** executes file "name" in this directory. This directory also contains the S executive.

doc:         Miscellaneous documents for use with S. Subdirectory **man** contains the manual page documentation for S (accessed by **S MAN**).

graph         Used for stand-alone graphics routines.

graph/bin:         Stand-alone graphics device drivers.

graph/lib:         Libraries for specific device drivers.

newfun         This branch of the hierarchy is necessary if users are to be able to create new S functions.

newfun/include:   Include files used by the **m4** macro processor. Provides facilities for the interface and algorithm languages. Accessed by the INCLUDE statement.

newfun/lib:         Libraries of algorithms.

s:         This directory contains the important portions of the S sytem.

s/.help:   All function detailed documentation.

s/data:   The shared data directory. Includes datasets and macros.

s/data/.help:   Detailed documentation for datasets and macros.

s/msg.d:   The "message of the day" for S users. Printed whenever S is started.

s/x:         All S functions that run as independent processes. This always includes graphics device drivers.

src         Source code for all of S.

src/fun:　　Contains many sub-directories; in general, one directory per S function. That directory contains all source code for that function and a makefile for constructing it. Also contains directories for each graphics device driver.

src/graph:　　Source code for the device-independent graphics algorithms.

src/icomp:　　Source for the S interface language compiler.

src/lang:　Basic S support algorithms.

src/main:　Source for the S executive: the parser and interpreter.

src/psl:　　Source for basic algorithms used by S for printing, storage allocation, numerical computations, etc. These routines can be used independently of S; see **S PROGRAM**.

src/util:　　Source for stand-alone S utilities, such as **extract** and **scandata**.

Note that some installations may not have certain of the S directories, either to conserve space or because of binary-only licenses.

The S home directory name differs depending on the local installation. The utility

　　S SHOME

will print the name of the S home directory on the standard output.

# Appendix D
# S Manual Pages

This appendix contains UNIX-style a manual page for S and a manual page for a set of C- or FORTRAN-callable subroutines that can create S datasets directly from a user program.

**NAME**

    S - Interactive Environment for Data Analysis and Graphics

**SYNOPSIS**

    S

**DESCRIPTION**

    *S* is a language for data analysis, graphics etc. Users type expressions to S; S evaluates the expressions. Results may be assigned to a permanent database or, if not, will be printed on the terminal. All expressions operate on self-describing data structures.

    Expressions may use standard arithmetic, logical and boolean operators, subsetting ([]) with arithmetic or logical expressions, component selection($) and various special operators. There are a large variety of functions for statistical, numerical and graphical techniques. Graphics may be done on any of a number of interactive terminals or on a printing terminal. The language can be extended by macros or user-written functions.

    On-line documentation is available through the *help* function.

**EXAMPLES**

```
1:10    # prints the integers 1 through 10
x <- rnorm(100)      # generate 100 pseudorandom numbers from
        # normal distribution, and save under the name x
regress(x,y)   # regression of y vector on x matrix
help("regress")   # print out detailed documentation for
                # regress function
mail #prompts for a letter to the S developers
```

**FILES**

```
swork    directory for working database
sdata    directory for database
sgraph   deferred graphics
sdump    file holds expression with syntax or execution error
sedit    file for editing of character vector or macro
Stemp*   temporary
```

**SEE ALSO**

    Richard A. Becker and John M. Chambers, *S: An Interactive Environment for Data Analysis and Graphics,* Wadsworth, 1984.

## NAME

    cvwrite - create S datasets directly from user programs

## SYNOPSIS

    From C:

    **cvint(name,data,n)**
    **char \*name; long data[]; long n;**

    **cvreal(name,data,n)**
    **char \*name; float data[]; long n;**

    **cvchar(name,data,n)**
    **char \*name; char \*data[]; long n;**

    **cmint(name,data,nrow,ncol)**
    **char \*name; long data[]; long nrow,ncol;**

    **cmreal(name,data,nrow,ncol)**
    **char \*name; float data[]; long nrow,ncol;**

    **cmchar(name,data,nrow,ncol)**
    **char \*name; char \*data[]; long nrow,ncol;**

    From Fortran:

    **subroutine cvectr(name,data,n,mode)**
    **CHARACTER(name,\*); POINTER data; integer n, mode**

    **subroutine cmatr(name,data,n,p,mode)**
    **CHARACTER(name,\*); POINTER data; integer n, p, mode**

## DESCRIPTION

    *cvint* creates an integer vector.

    *cvreal* creates a real vector.

    *cvchar* creates a character string vector.

    *cmint* creates an integer matrix.

    *cmreal* creates a real matrix.

    *cmchar* creates a character string matrix.

    *cvectr* creates a vector of length n.

    *cmatr* creates an n by p matrix.

    The values of *mode* are 2=integer, 3=real, 5=character.

    Character data uses the S algorithm language conventions that a vector of character strings is actually a vector of pointers (integers), each of which points to a (null terminated) character string on the stack.

    Appropriate libraries must be searched to get the routines:
        LIBDIR='S SHOME'/newfun/lib     #directory containing libraries
        f77 ... object files ... $LIBDIR/lang $LIBDIR/psl
            # load to create a.out file

    Note that loading must be accomplished by the *f77* command, and not by *cc* . This is to ensure that Fortran system libraries are searched. Also, you should have the following line in your C main program:
        MAIN__0{}
    to define a dummy routine named MAIN__. F77 thinks it is needed.

    Once your a.out file is created, it should be run inside the swork or sdata directory where the datasets should be placed, or the created files should be moved to an S data directory afterward.

# Appendix E
# Maintaining S

This appendix contains information of use to the person designated to maintain S on a computer system.

# Maintaining S

These brief instructions give miscellaneous information to the local maintainer of S that cannot be found in the elsewhere. This information should be of value in the (unlikely?) circumstance that portions of S need alterations.

## 1. The S Directory Structure

S makes extensive use of the UNIX† hierarchical file system. At the top level, the code is split into directories

| | |
|---|---|
| adm | administrator commands |
| cmd | S shell scripts and executables |
| doc | various pieces of S documentation |
| graph | allows use of S standalone graphics |
| newfun | allows user-written function |
| s | the executable code for S functions |
| | includes system database, online documentation |
| src | all source code for S |

Of these, only the "s" and "cmd" directories are absolutely crucial to the running of S. Other directories can be removed to save disk space with the corresponding loss of functionality. The organization of the "src" directory is perhaps the most crucial to modifying and maintaining S:

| | |
|---|---|
| fun | all S functions, one (or a small set) per directory ($F) |
| graph | graphical support algorithms ($GRZ) |
| icomp | interface language compiler (probably best left alone) |
| lang | S support algorithms ($S) |
| main | S executive |
| psl | general algorithms ($P) |
| util | S standalone utilities |
| yacc | Ratfor-producing version of yacc |

The ENVIRONMENT file in the S home directory defines a number of shell variables that are useful in locating particular parts of the hierarchy without having to follow through the various levels. For example, the following variables contain the names of directories with key portions of the S code:

---

† UNIX is a Trademark of AT&T Bell Laboratories.

| | |
|---|---|
| $F | functions |
| $M | executive |
| $I | INCLUDE files |
| $L | libraries |
| $P | general support programs (creates $L/psl) |
| $S | S support programs (creates $L/lang) |
| $GRZ | graphical subroutines (creates $L/grz) |
| $C | utilities (invoked as S *file*) |
| $A | utilities for the S maintainer (i.e. you) |
| $HELP | detailed function documentation |

It is common to put the line

. ENVIRONMENT

into the .profile file for S so that these variables are available whenever you are logged in as "s".

## 2. Modifying Underlying Algorithms

In general, the libraries are updated by changing the source code for the selected programs, using S MAKE to create ".o" versions of the source files, and using *ar* (and *ranlib* if you have it) to maintain the library archive. For example,

```
$ cd $P
$ ed mean.r
   ... make changes with editor ...
$ S MAKE mean.o
$ ar r $L/psl mean.o
$ ranlib $L/psl              # Berkeley UNIX only
```

Of course, you must re-load other functions that search the library before the changed routines are actually used by S.

## 3. Modifications to S functions

S functions may be loaded as part of the S executive or as separate processes. The major advantages of the separate process form is robustness to problems and smaller main-memory requirements; the advantage of the internal form is higher execution speed and smaller disk usage. In general, on machines with lots of main memory, the more functions in the executive, the better. The file $M/infun.list controls which functions are to be internal to the S executive. If you modify this file, you should execute

$ $A/INTERNAL

which recreates $M/ILIST.a, a long process, after which you must load a new S executive (section 4).

## 3.1.  External Functions

The procedure for modifying an S function differs depending on whether the function is internal or external.  To modify an external S function, get into the appropriate directory ($F/fun) modify the source code (and Smakefile if files were added or deleted), and type

```
$ S MAKE function
```

This will compile and load the function.  To test it, copy it to your test directory (for example, /usr/joe/x, all test directories must be named "x") as follows:

```
$ cp fun /usr/joe/x
$ S
> chapter("/usr/joe")
```

will allow you to test the new version.  Once it has been tested,

```
$ cd $F/fun
$ $A/INSTALL fun
```

will install it into the generally available S functions.  In case of disaster,

```
$ cd $F/fun
$ $A/BACKUP fun
```

will bring the old version back.

## 3.2.  Internal Functions

To change an internal function, modify the appropriate code and then test it as if it were an external function:

```
$ cd $F/fun
... change things ...
$ S MAKE
$ cp fun /usr/joe/x/testfun
$ S
> chapter("/usr/joe")
> testfun(...)  # try the function
```

When you are convinced that your modified code is ok, put the object code corresponding to your function on $M/ILIST.a.  This provides a convenient way of reloading the executive without always picking up developmental function code.

```
$ ar r $M/ILIST.a files
$ ranlib $M/ILIST.a
$ cd $M
$ S MAKE NEW.S     # load a new executive
```

Now test the new executive by using

```
$ S NEW.S
```

instead of simply "S". Once the new function (and executive) is working properly,

```
$ cd $M
$ S MAKE INSTALL
```

## 4. Modifying the S Executive

If you want to change parts of the S executive,

```
$ cd $M
... change files ...
$ S MAKE NEW.S     # for any changed files
```

Test it, and then

```
$ S MAKE INSTALL     # installs new executive
```

The command,

```
$ S MAKE BACKUP
```

is available in case of problems; it re-installs the old executive "$C/OLD.S".

A set of test files is provided to exercise many of the basic S functions. Also included is the anticipated output of these tests. The tests are named *arith*, *basic*, *basic2*, *large*, *prob*, and *apply*. They are run (in background mode) by typing

```
$ $A/DOTEST name
```

You can run all tests by using

```
$ $A/DOTEST ALL
```

To test a new version of the executive before installing

```
$ $A/DOTEST NEW.S ALL
```

The test results are placed in files $TEST/current/*name*, and this can be compared with files $TEST/target/*name*. The files $TEST/plot and $TEST/device can be used as source files to exercise the graphical functions.

**5. Installation of S**

One task that may fall to a local S maintainer: if S cannot be given the login id "s", then two things need to be edited. First, *$C/NEWUSER* sends mail to "s" whenever a new S user is processed. This file can be changed to send mail anywhere desired (or the mail can be eliminated). Also, the *mail* function needs changes. Edit the file *$F/mail/mail.i* to change the mail commands contained therein, reload it as an internal or external function.

The file

$GUIDE/man/S.1

is in the proper format as a manual page for the S system. It should be installed somewhere where the "man" command can get at it. Another mechanism for communicating with local S users is by means of the file $SHOME/s/msg.d. The contents of the file (normally empty) are printed each time S is run.

**6. Miscellaneous Stuff**

Another potentially interesting utility is *LISTALL*; this provides a means of making a neatly formatted listing of all source files in one of the S directories listed above. Get into the directory, and invoke the utility:

```
$ cd $P
$ nohup $A/LISTALL &     # offline listing of $P source files
```

(The shell script $A/LISTALL may need to be changed to replace command *lpr* with the local off-line print command.)

If S is run in a directory containing a file named "S.profile", the names of all functions used during execution will be appended to the file.

```
$ touch S.profile
$ S
   ... execute S expressions
> q
$ S SUMMARY
```

The utility "S SUMMARY" produces a count of the number of times each function was executed. Functions which are heavily used may be candidates for inclusion as part of the S executive.

# Index

[ operator  10
#, comment  123
%, special  operator  122
& argument  continuation  29
   CHAIN  continuation  34
   ENDARGS  suppressing  79
   RETURN  continuation  34
? macro  invocation  122
ABORT statement  43, 47
actual argument  28
adding to plots  82
adj graphical  parameter  75, 87
algebra algorithms,  linear  65
algorithm facility  46
        facility,  attach  55
            graphics  71
            io  49
            print  47
            read  52
            stack  56
            struct  57
            tree  58
      language  2, 3, 41, 71
algorithm, cfill  60
    ch2vec  61
    chcomp  62
    concat  60

icopy  60
ifill  60
istrng  50, 61
jgetch  60
jputch  60
jstkgt  56
jstkrl  56
lcopy  60
lfill  60
match  19
narang  63, 72
orderf  62
orderi  62
orderr  62
rangec  62, 72
rangev  62, 72
rcopy  60
rdtfmt  51
rfill  60
sattac  55
sclose  55
sdetac  55
sopen  55
sort  61
sortfr  61
sorti  61
streq  61

algorithms 5
algorithms, graphical  67
      linear  algebra  65
      probability  63
      quantile  63
      random  number  63
ALLARG statement  30
allocation, dynamic  storage  56
      scratch  space  9
alphanumeric mode  101
ANY specifier  24
arbitrarily many  arguments  30
ARG statement  29
ARGSTR statement  33
argument 126
      continuation, &  29
      list  from  structure  32
      partial  match  28, 128
      processing  127
      processing,  interrupting
         29
      structure  127
argument, actual  28
      FILTER  35
      formal  28
      MISSING  7
      missing  128
      mode  of  6
      name  of  28
      named  30
      OPTIONAL  7
      optional  7
      PAR  35, 38
      PLOTARGS  39
      type  of  6
      USED  33
arguments to functions  5
arguments, arbitrarily  many  30
      avoiding  named  31
      default  values  for  7
      looping  over  30
arowsz graphical  subroutine  73
arpltz graphical  subroutine  85
assign function  130, 134

attach algorithm  facility  55
attribute 10
attribute, DATANAME  20
      ENTRY  58
      FIRSTENT  58
      FNAME  20
      KEYNAME  20
      LASTENT  58
      LENGTH  10, 18, 57
      MISSING  10, 28
      MODE  18, 57
      NAME  57
      NCOL  10, 18
      NROW  10, 18
      STRING  19
      TEND  10, 18
      TEXT  18
      TNPER  10, 18
      TSTART  10, 18
      VALUE  27, 57
AUTO_CHECK variable  54
avoiding named  arguments  31
axes function  38, 80
axis labels (pretty)  90
      numeric  labels  74
      parameters  91
AXIS statement  72, 89
axis style  90, 91
      parameter  90
      tick  marks  74
      type  91
axis, logarithmic  90
      time  90
barz graphical  subroutine  95
beginz graphical  subroutine  71
bhtchz graphical  subroutine  96
BIG constant  44
BIGEXP constant  44
bmglgz graphical  subroutine  96
bnamez graphical  subroutine  96
BOTTOM side  of  plot  74
bplotz graphical  subroutine  95
break syntax  132
BUFFER variable  49

BUFLEN variable  49
BUFPOS variable  49
built-in dataset  132
     mode  25
bxnz graphical  subroutine  96
bxpz graphical  subroutine  95
bxz graphical  subroutine  96
C format  item  48, 53
  language  45
  routines  with  functions  12
CASE statement  24
cex graphical  parameter  76
cfill algorithm  60
ch2vec algorithm  61
CHAIN continuation, &  34
     statement  36, 80
changing graphical  parameters  70
chapter function  13, 127
CHAPTER utility  11
CHAR mode  44
character data,  pointer  to  18
CHARACTER statement  44
character string  122
        data  18, 50, 60
character, EOS  18, 61
chcomp algorithm  62
Choleski decomposition  65
clipping 87
closing, file  55
co-ordinate systems  for  graphical
     algorithms  72
coerce 6
COERCE statement  26, 31
col graphical  parameter  76
COMMA format  item  52
command 125
command, pwd  13
    sdb  16
comment #  123
common block  declaration  9
compilation 13
component insertion,  data  struc-
     ture  37
    select  21

components with  same  names  22
compound expression  125, 131
    expression,  value  of
     131
concat algorithm  60
conditional expression  132
consistency of  graphical  parame-
    ters  85
constant, BIG  44
    BIGEXP  44
    DEG2RD  44
    EOS  44, 50
    ESCAPE  44
    FIELDWIDTH  49
    LARGEINT  44
    LINEWIDTH  49
    NBPC  44
    NCPW  44
    NDIGITS  44
    PI  44
    PRECISION  44
    SMALL  44
cont2z graphical  subroutine  96
continuation 125
contrz graphical  subroutine  95
coordinate system,  device  97
     margin  69, 86
     raster  97
     user  69, 89, 91
COPY specifier  22
crclsz graphical  subroutine  73
crt graphical  parameter  76
csi graphical  parameter  76
csr graphical  parameter  76
CTABLE statement  19
cxy graphical  parameter  75
cyclic use of  data  24
data structure  57, 133
     component  insertion
     37
   structure,  depth-first  scan  of
     58
     dynamically  allo-
     cated  8

external representation of 134
internal representation of 134
NA in 21
plot 39
data, character string 18, 50, 60
cyclic use of 24
DATANAME attribute 20
dataset, built-in 132
Random.seed 64
DEBUG statement 14, 45
debugging, UNIX 16
declaration, common block 9
declarations, FORTRAN 9
DECODE statement 53
decoding facilities 52
decomposition, Choleski 65
q-r 65
singular-value 65
default labels 20
values for arguments 7
DEG2RD constant 44
depth-first scan of data structure 58
description list 6
descriptor, file 55
design of functions 5
device coordinate system 97
driver, graphical 82, 96
DEVICE_DRIVER statement 103
dget function 133, 134
din graphical parameter 88
diraxz graphical subroutine 91
documentation, function 13
dput function 133, 134
drfigz graphical subroutine 85
driver, graphical device 82
drpltz graphical subroutine 85
dump function 133
dumpaz graphical subroutine 94
dynamic storage allocation 56
dynamically allocated data structure 8

EDITDOC utility 14
eepltz graphical subroutine 95
eigenvalue 65
encode example 23
ENCODE statement 23, 49
end of file 54
ENDARGS statement 29
suppressing, & 79
ENDARGS, suppression 29
ENTRY attribute 58
environment, S 41
EOF variable 54
EOS character 18, 61
constant 44, 50
EPRINT statement 47
eqpltz graphical subroutine 95
error message 7, 43
ESCAPE constant 44
example, encode 23
plotting 77, 79
pmax 25
expression 123
expression, compound 125, 131
conditional 132
iterative 132
value of compound 131
external representation of data structure 134
extract macro 134
utility 53, 134
facilities, decoding 52
reading 52
FATAL statement 8, 43
FIELDWIDTH constant 49
figure type, graphical 85
figure, graphical 68, 84
file 134
closing 55
descriptor 55
name suffix 12, 41
opening 55
file, end of 54
FILTER argument 35

fin graphical parameter 88
finisz graphical subroutine 77
FIRSTENT attribute 58
FNAME attribute 20
for syntax 132
formal argument 28
format item, C 48, 53
    COMMA 52
    I 48, 53
    L 48, 53
    parenthesis 52
    Q 48
    R 48, 53
    S 53
    SHARP 52
    SP 50
    T 50
    TM 50
FORTRAN declarations 9
    language 4, 41
FPRINT statement 48
FROM specifier 22, 32
fty graphical parameter 85
function call 126
    chaining 130
    documentation 13
FUNCTION statement 6
    utility 12, 37, 42, 64, 103
function values 34
function, assign 130, 134
    axes 38, 80
    chapter 13, 127
    dget 133, 134
    dput 133, 134
    dump 133
    get 134
    macro 122
    options 130
    pardump 94
    read 134
    restore 133
    save 130
    search 13

write 134
functions, arguments to 5
    C routines with 12
    design of 5
    graphics 37
    high-level graphics 39
    libraries used by 12
    name of 12
    new 3
    object files used by 12
    options to 5
    results of 5
FUNCTON utility 78
generators, pseudorandom 63
get function 134
GETLINE statement 54
GETSEED statement 64
GETSTRING statement 55
graph mode 101
graphical algorithms 67
    algorithms, co-ordinate systems for 72
        labeling in 74
        plotting data in 73
    device driver 82, 96
    figure 68, 84
        type 85
    input 93, 102
    margins 68, 84
    outer margins 68, 83
    parameter, adj 75, 87
        cex 76
        col 76
        crt 76
        csi 76
        csr 76
        cxy 75
        din 88
        fin 88
        fty 85
        lab 89
        las 89

lty 76
lwd 76
mai 84
mar 84
mex 84, 85, 86
mfg 85
mgp 89
oma 85
omi 85
omo 87
PAR 79
pin 88
pty 84
srt 75
tck 89
uin 88
usr 89, 91
xaxp 91
xaxs 91
xaxt 91
xpd 87
yaxp 91
yaxs 91
yaxt 91
parameters 70, 74
PAR 35, 37
parameters, changing 70
consistency
of 85
plot 68, 84
type 84
subroutine, arowsz 73
arpltz 85
barz 95
beginz 71
bhtchz 96
bmglgz 96
bnamez 96
bplotz 95
bxnz 96
bxpz 95
bxz 96
cont2z 96

contrz 95
crclsz 73
diraxz 91
drfigz 85
drpltz 85
dumpaz 94
eepltz 95
eqpltz 95
finisz 77
gtextz 87
hatch 96
hdlinz 96
hhbrkz 96
hhdonz 96
hhpltz 95
i6to9z 102
inpltz 85
inputz 93
laxisz 91
linesp 96
linesz 73
lintfz 96
logaxz 90
mfcolz 85
mfrowz 85
mtextz 74, 87
nplotz 95
npntsz 96
otextz 87
persp 95
pictur 96
piez 95
plot1 95
pltusa 95
pnttfz 96
pointz 73
ptitle 96
rplotz 95
rpntsz 96
saxisz 74, 91
segmtz 73
shadez 96
splotz 95
stdaxz 90

taxisz 91
textz 73
timaxz 90
titlez 96
tspltz 95
tsvecz 96
usabdy 96
zejecz 101
zlinez 101
zoutrz 102
zoutwz 102
zparmz 100
zptchz 101
zquxyz 102
zseekz 100
zwrapz 101
title 74
graphics algorithm facility 71
functions 37
functions, high-level 39
NA in 39
gtextz graphical subroutine 87
hatch graphical subroutine 96
hdlinz graphical subroutine 96
hhbrkz graphical subroutine 96
hhdonz graphical subroutine 96
hhpltz graphical subroutine 95
hierarchical structure 21
high-level graphics functions 39
I format item 48, 53
i6to9z graphical subroutine 102
icopy algorithm 60
ifill algorithm 60
INBUF variable 52
INCLUDE statement 46
INLEN variable 52
inpltz graphical subroutine 85
INPOS variable 52
input, graphical 93, 102
inputz graphical subroutine 93
INSERT statement 37
insertion, data structure component 37
INT mode 44

interface language 1, 3
subset 10
language, operating on
NA in 21
internal representation of data
structure 134
interpreter 121
interrupting argument processing
29
io algorithm facility 49
istrng algorithm 50, 61
iterative expression 132
jgetch algorithm 60
jputch algorithm 60
jstkgt algorithm 56
jstkrl algorithm 56
KEYNAME attribute 20
L format item 48, 53
lab graphical parameter 89
labeling in graphical algorithms
74
labels, default 20
language, algorithm 2, 3, 41, 71
C 45
FORTRAN 4, 41
interface 1, 3
RATFOR 4, 10, 41
LARGEINT constant 44
las graphical parameter 89
LASTENT attribute 58
laxisz graphical subroutine 91
lcopy algorithm 60
LEFT side of plot 74
LENGTH attribute 10, 18, 57
lexical analyzer 121
lfill algorithm 60
libraries used by functions 12
LIKE specifier 9
linear algebra algorithms 65
linesp graphical subroutine 96
linesz graphical subroutine 73
LINEWIDTH constant 49
lintfz graphical subroutine 96
list, description 6

logarithmic axis 90

        plot 40

logaxz graphical subroutine 90

LOGPLOT statement 40

looping over arguments 30

lty graphical parameter 76

lwd graphical parameter 76

m4 macro processor 12, 58

macro 3

    function 122

    invocation, ? 122

    processor, m4 58

macro, extract 134

mai graphical parameter 84

MAKE utility 13, 43, 103

mar graphical parameter 84

margin coordinate system 69, 86

margins, graphical 68, 84

        outer 68

match algorithm 19

match, argument partial 28

    partial 20

MESSAGE statement 43

message, error 7, 43

methods, user-supplied 19

mex graphical parameter 84, 85, 86

mfcolz graphical subroutine 85

mfg graphical parameter 85

mfrowz graphical subroutine 85

mgp graphical parameter 89

MISSING argument 7

missing argument 128

MISSING attribute 10, 28

missing values (see NA) 21

MODE attribute 18, 57

mode of argument 6

    specifier 6

mode, alphanumeric 101

    built-in 25

    CHAR 44

    determined by position 26

    graph 101

    INT 44

    REAL 44

MODECALC specifier 25

mtextz graphical subroutine 74, 87

NA 63

    in data structure 21

        graphics functions 39

        interface language, operating on 21

    strategy 21

NA, set 21

    test for 21

name 122

NAME attribute 57

name of argument 28

        functions 12

      suffix, file 12, 41

name, reserved 126

named argument 30

names, components with same 22

NAOK specifier 21

narang algorithm 63, 72

NBPC constant 44

NCOL attribute 10, 18

NCPW constant 44

NDIGITS constant 44

NEWDOC utility 14

next syntax 132

NEXTARG statement 30

NOKEY specifier 31

NOPRINT statement 35

nplotz graphical subroutine 95

npntsz graphical subroutine 96

NROW attribute 10, 18

numeric labels, axis 74

object files used by functions 12

oma graphical parameter 85

omi graphical parameter 85

omo graphical parameter 87

opening, file 55

operating on NA in interface language 21

operator 129

    %, special 122

operator, [ 10

OPTIONAL argument 7

optional argument 7
OPTIONAL specifier 28
options function 130
   to functions 5
orderf algorithm 62
orderi algorithm 62
orderr algorithm 62
otextz graphical subroutine 87
outer margins, graphical 68, 83
PAR argument 35, 38
  graphical parameter 79
PAR, graphical parameters 35, 37
parameter, axis style 90
parameters PAR, graphical 35, 37
parameters, axis 91
   changing graphical 70
   graphical 70, 74
pardump function 94
parenthesis format item 52
parser 121
partial match 20
  match, argument 28
persp graphical subroutine 95
PI constant 44
pictur graphical subroutine 96
piez graphical subroutine 95
pin graphical parameter 88
plot data structure 39
  type, graphical 84
plot, graphical 68, 84
  logarithmic 40
plot1 graphical subroutine 95
PLOTARGS argument 39
   statement 79
plots, adding to 82
plotting data in graphical algo-
    rithms 73
   example 77, 79
plotting, axis style 90
pltusa graphical subroutine 95
pmax example 25
pnttfz graphical subroutine 96
pointer 56
POINTER statement 44

pointer to character data 18
pointz graphical subroutine 73
PRECISION constant 44
pretty numbers for axis labels 90
print algorithm facility 47
PRINT statement 47
print, automatic 131
PRINTDOC utility 14
probability algorithms 63
PROGRAM utility 42
PROMPT utility 14
pseudorandom generators 63
ptitle graphical subroutine 96
pty graphical parameter 84
PUTSEED statement 64
pwd command 13
Q format item 48
q-r decomposition 65
quantile algorithms 63
QUERY statement 76, 93
R format item 48, 53
random number algorithms 63
Random.seed dataset 64
rangec algorithm 62, 72
rangev algorithm 62, 72
raster coordinate system 97
RATFOR language 4, 10, 41
rcopy algorithm 60
rdtfmt algorithm 51
read algorithm facility 52
  function 134
READ statement 52
reading facilities 52
REAL mode 44
renaming in RETURN 35
repeat syntax 132
reserved name 126
restore function 133
results of functions 5
RETURN continuation, & 34
   statement 11, 34
RETURN, how it works 35
   renaming in 35
rfill algorithm 60

RIGHT side of plot 74
rplotz graphical subroutine 95
rpntsz graphical subroutine 96
S environment 41
  format item 53
sattac algorithm 55
save function 130
saxisz graphical subroutine 74, 91
sclose algorithm 55
scratch space allocation 9
sdb command 16
sdetac algorithm 55
search function 13
segmtz graphical subroutine 73
select, component 21
semantics 126
set NA 21
SETUP statement 39, 79
shadez graphical subroutine 96
SHARP format item 52
singular-value decomposition 65
SMALL constant 44
sopen algorithm 55
sort algorithm 61
sortfr algorithm 61
sorti algorithm 61
SP format item 50
special operator % 122
specifier, ANY 24
        COPY 22
        FROM 22, 32
        LIKE 9
        mode 6
        MODECALC 25
        NAOK 21
        NOKEY 31
        OPTIONAL 28
        STR 21
        TSTRING 61
        type 6
SPECIFY statement 76, 93
splotz graphical subroutine 95
srt graphical parameter 75
stack algorithm facility 56

statement, ABORT 43, 47
        ALLARG 30
        ARG 29
        ARGSTR 33
        AXIS 72, 89
        CASE 24
        CHAIN 36, 80
        CHARACTER 44
        COERCE 26, 31
        CTABLE 19
        DEBUG 14, 45
        DECODE 53
        DEVICE_DRIVER 103
        ENCODE 23, 49
        ENDARGS 29
        EPRINT 47
        FATAL 8, 43
        FPRINT 48
        FUNCTION 6
        GETLINE 54
        GETSEED 64
        GETSTRING 55
        INCLUDE 46
        INSERT 37
        LOGPLOT 40
        MESSAGE 43
        NEXTARG 30
        NOPRINT 35
        PLOTARGS 79
        POINTER 44
        PRINT 47
        PUTSEED 64
        QUERY 76, 93
        READ 52
        RETURN 11, 34
        SETUP 39, 79
        SPECIFY 76, 93
        STATIC 9
        STRUCTURE 8
        SWITCH MODE 24
        TREEWALK 59
        WARNING 43
STATIC statement 9
stdaxz graphical subroutine 90

storage allocation, dynamic  56
STR specifier  21
streq algorithm  61
STRING attribute  19
string data, character  18, 50
string, character  122
struct algorithm  facility  57
STRUCTURE statement  8
structure, argument  127
         list  from  32
     data  57, 133
     depth-first  scan  of  data
       58
     dynamically  allocated
       data  8
     external  representation
       of data  134
     hierarchical  21
     internal  representation
       of data  134
     NA  in  data  21
style, axis  90, 91
subset, interface  language  10
suffix, file  name  12, 41
SWITCH MODE  statement  24
syntax rule  121
syntax, break  132
     for  132
     next  132
     repeat  132
     while  132
T format  item  50
taxisz graphical  subroutine  91
tck graphical  parameter  89
TEND attribute  10, 18
test for  NA  21
TEXT attribute  18
textz graphical  subroutine  73
tick marks,  axis  74
timaxz graphical  subroutine  90
time axis  90
title, graphical  74
titlez graphical  subroutine  96
TM format  item  50

TNPER attribute  10, 18
token 121
TOP side  of  plot  74
tree algorithm  facility  58
TREEWALK statement  59
tspltz graphical  subroutine  95
TSTART attribute  10, 18
TSTRING specifier  61
tsvecz graphical  subroutine  96
type of  argument  6
     specifier  6
type, axis  91
uin graphical  parameter  88
UNIX debugging  16
usabdy graphical  subroutine  96
USED argument  33
     value  20
user coordinate  system  69, 89, 91
user-supplied methods  19
usr graphical  parameter  89, 91
utility, CHAPTER  11
     EDITDOC  14
     extract  53, 134
     FUNCTION  12, 37, 42, 64,
       103
     FUNCTON  78
     MAKE  13, 43, 103
     NEWDOC  14
     PRINTDOC  14
     PROGRAM  42
     PROMPT  14
VALUE attribute  27, 57
value of  compound  expression
     131
value, USED  20
values, function  34
variable, AUTO_CHECK  54
     BUFFER  49
     BUFLEN  49
     BUFPOS  49
     EOF  54
     INBUF  52
     INLEN  52
     INPOS  52

WARNING statement  43
while syntax  132
write function  134
xaxp graphical  parameter  91
xaxs graphical  parameter  91
xaxt graphical  parameter  91
xpd graphical  parameter  87
yaxp graphical  parameter  91
yaxs graphical  parameter  91
yaxt graphical  parameter  91

zejecz graphical  subroutine  101
zlinez graphical  subroutine  101
zoutrz graphical  subroutine  102
zoutwz graphical  subroutine  102
zparmz graphical  subroutine  100
zptchz graphical  subroutine  101
zquxyz graphical  subroutine  102
zseekz graphical  subroutine  100
zwrapz graphical  subroutine  101